On War.txt

ON WAR

by General Carl von Clausewitz

ON WAR GENERAL CARL VON CLAUSEWITZ TRANSLATED BY COLONEL J.J. GRAHAM

1874 was 1st edition of this translation. 1909 was the London reprinting.

NEW AND REVISED EDITION WITH AN INTRODUCTION AND NOTES BY COLONEL F.N. MAUDE C.B. (LATE R.E.)

EIGHTH IMPRESSION IN THREE VOLUMES

VOLUME I

INTRODUCTION

THE Germans interpret their new national colours--black, red, and white--by the saying, "Durch Nacht und Blut zur licht." ("Through night and blood to light"), and no work yet written conveys to the thinker a clearer conception of all that the red streak in their flag stands for than this deep and philosophical analysis of "War" by Clausewitz.

It reveals "War," stripped of all accessories, as the exercise of force for the attainment of a political object, unrestrained by any law save that of expediency, and thus gives the key to the interpretation of German political aims, past, present, and future, which is unconditionally necessary for every student of the modern conditions of Europe. Step by step, every event since Waterloo follows with logical consistency from the teachings of Napoleon, formulated for the first time, some twenty years afterwards, by this remarkable thinker.

What Darwin accomplished for Biology generally Clausewitz did for the Life-History of Nations nearly half a century before him, for both have proved the existence of the same law in each case, viz., "The survival of the fittest"--the "fittest," as Huxley long since pointed out, not being necessarily synonymous with the ethically "best." Neither of these thinkers was concerned with the ethics of the struggle which each studied so exhaustively, but to both men the phase or condition presented itself neither as moral nor immoral, any more than are famine, disease, or other natural phenomena, but as emanating from a force inherent in all living organisms which can only be mastered by understanding its nature. It is in that spirit that, one after the other, all the Nations of the Continent, taught by such drastic lessons as Koniggraetz and Sedan, have accepted the lesson, with the result that to-day Europe is an armed camp, and peace is maintained by the equilibrium of forces, and will continue just as long as this equilibrium exists, and no longer.

Whether this state of equilibrium is in itself a good or desirable thing may be open to argument. I have discussed it at length in my "War and the World's Life"; but I venture to suggest that to no one would a renewal of the era of warfare be a change for the better, as far as existing humanity is concerned. Meanwhile, however, with every year that elapses the forces at present in equilibrium are changing in magnitude--the pressure of populations which have to be fed is rising, and an explosion along the line of least resistance is, sooner or later, inevitable.

As I read the teaching of the recent Hague Conference, no responsible
Government on the Continent is anxious to form in themselves that line
of least resistance; they know only too well what War would mean; and
we alone, absolutely unconscious of the trend of the dominant thought
of Europe, are pulling down the dam which may at any moment let in on us
the flood of invasion.

Now no responsible man in Europe, perhaps least of all in Germany,
thanks us for this voluntary destruction of our defences, for all who
are of any importance would very much rather end their days in peace
than incur the burden of responsibility which War would entail. But
they realise that the gradual dissemination of the principles taught by
Clausewitz has created a condition of molecular tension in the minds of
the Nations they govern analogous to the "critical temperature of water
heated above boiling-point under pressure," which may at any moment
bring about an explosion which they will be powerless to control.

The case is identical with that of an ordinary steam boiler, delivering
so and so many pounds of steam to its engines as long as the
envelope can contain the pressure; but let a breach in its continuity
arise--relieving the boiling water of all restraint--and in a moment the
whole mass flashes into vapour, developing a power no work of man can
oppose.

The ultimate consequences of defeat no man can foretell. The only way to
avert them is to ensure victory; and, again following out the principles
of Clausewitz, victory can only be ensured by the creation in peace of
an organisation which will bring every available man, horse, and gun (or
ship and gun, if the war be on the sea) in the shortest possible time,
and with the utmost possible momentum, upon the decisive field of
action--which in turn leads to the final doctrine formulated by Von der
Goltz in excuse for the action of the late President Kruger in 1899:

"The Statesman who, knowing his instrument to be ready, and seeing War
inevitable, hesitates to strike first is guilty of a crime against his
country."

It is because this sequence of cause and effect is absolutely unknown to
our Members of Parliament, elected by popular representation, that
all our efforts to ensure a lasting peace by securing efficiency with
economy in our National Defences have been rendered nugatory.

This estimate of the influence of Clausewitz's sentiments on
contemporary thought in Continental Europe may appear exaggerated to
those who have not familiarised themselves with M. Gustav de Bon's
exposition of the laws governing the formation and conduct of crowds I
do not wish for one minute to be understood as asserting that Clausewitz
has been conscientiously studied and understood in any Army, not even
in the Prussian, but his work has been the ultimate foundation on which
every drill regulation in Europe, except our own, has been reared. It is
this ceaseless repetition of his fundamental ideas to which one-half of
the male population of every Continental Nation has been subjected
for two to three years of their lives, which has tuned their minds to
vibrate in harmony with his precepts, and those who know and appreciate
this fact at its true value have only to strike the necessary chords
in order to evoke a response sufficient to overpower any other ethical
conception which those who have not organised their forces beforehand
can appeal to.

The recent set-back experienced by the Socialists in Germany is an
illustration of my position. The Socialist leaders of that country
are far behind the responsible Governors in their knowledge of the
management of crowds. The latter had long before (in 1893, in fact)
made their arrangements to prevent the spread of Socialistic propaganda
beyond certain useful limits. As long as the Socialists only threatened
capital they were not seriously interfered with, for the Government

knew quite well that the undisputed sway of the employer was not for the
ultimate good of the State. The standard of comfort must not be pitched
too low if men are to be ready to die for their country. But the moment
the Socialists began to interfere seriously with the discipline of the
Army the word went round, and the Socialists lost heavily at the polls.

If this power of predetermined reaction to acquired ideas can be
evoked successfully in a matter of internal interest only, in which the
"obvious interest" of the vast majority of the population is so clearly
on the side of the Socialist, it must be evident how enormously greater
it will prove when set in motion against an external enemy, where the
"obvious interest" of the people is, from the very nature of things, as
manifestly on the side of the Government; and the Statesman who failed
to take into account the force of the "resultant thought wave" of a
crowd of some seven million men, all trained to respond to their ruler's
call, would be guilty of treachery as grave as one who failed to strike
when he knew the Army to be ready for immediate action.

As already pointed out, it is to the spread of Clausewitz's ideas that
the present state of more or less immediate readiness for war of all
European Armies is due, and since the organisation of these forces is
uniform this "more or less" of readiness exists in precise proportion to
the sense of duty which animates the several Armies. Where the spirit of
duty and self-sacrifice is low the troops are unready and inefficient;
where, as in Prussia, these qualities, by the training of a whole
century, have become instinctive, troops really are ready to the last
button, and might be poured down upon any one of her neighbours with
such rapidity that the very first collision must suffice to ensure
ultimate success--a success by no means certain if the enemy, whoever he
may be, is allowed breathing-time in which to set his house in order.

An example will make this clearer. In 1887 Germany was on the very verge
of War with France and Russia. At that moment her superior efficiency,
the consequence of this inborn sense of duty--surely one of the highest
qualities of humanity--was so great that it is more than probable that
less than six weeks would have sufficed to bring the French to their
knees. Indeed, after the first fortnight it would have been possible
to begin transferring troops from the Rhine to the Niemen; and the same
case may arise again. But if France and Russia had been allowed even
ten days' warning the German plan would have been completely defeated.
France alone might then have claimed all the efforts that Germany could
have put forth to defeat her.

Yet there are politicians in England so grossly ignorant of the German
reading of the Napoleonic lessons that they expect that Nation to
sacrifice the enormous advantage they have prepared by a whole century
of self-sacrifice and practical patriotism by an appeal to a Court of
Arbitration, and the further delays which must arise by going through
the medieaeval formalities of recalling Ambassadors and exchanging
ultimatums.

Most of our present-day politicians have made their money in business--a
"form of human competition greatly resembling War," to paraphrase
Clausewitz. Did they, when in the throes of such competition, send
formal notice to their rivals of their plans to get the better of them
in commerce? Did Mr. Carnegie, the arch-priest of Peace at any price,
when he built up the Steel Trust, notify his competitors when and how
he proposed to strike the blows which successively made him master
of millions? Surely the Directors of a Great Nation may consider the
interests of their shareholders--i.e., the people they govern--as
sufficiently serious not to be endangered by the deliberate sacrifice
of the preponderant position of readiness which generations of
self-devotion, patriotism and wise forethought have won for them?

As regards the strictly military side of this work, though the recent
researches of the French General Staff into the records and documents of
the Napoleonic period have shown conclusively that Clausewitz had never
grasped the essential point of the Great Emperor's strategic method,

yet it is admitted that he has completely fathomed the spirit which gave
life to the form; and notwithstandingthe variations in application which
have resulted from the progress of invention in every field of national
activity (not in the technical improvements in armament alone), this
spirit still remains the essential factor in the whole matter. Indeed,
if anything, modern appliances have intensified its importance, for
though, with equal armaments on both sides, the form of battles must
always remain the same, the facility and certainty of combination which
better methods of communicating orders and intelligence have
conferred upon the Commanders has rendered the control of great masses
immeasurably more certain than it was in the past.

Men kill each other at greater distances, it is true--but killing is a
constant factor in all battles. The difference between "now and then"
lies in this, that, thanks to the enormous increase in range (the
essential feature in modern armaments), it is possible to concentrate
by surprise, on any chosen spot, a man-killing power fully twentyfold
greater than was conceivable in the days of Waterloo; and whereas in
Napoleon's time this concentration of man-killing power (which in his
hands took the form of the great case-shot attack) depended almost
entirely on the shape and condition of the ground, which might or might
not be favourable, nowadays such concentration of fire-power is almost
independent of the country altogether.

Thus, at Waterloo, Napoleon was compelled to wait till the ground became
firm enough for his guns to gallop over; nowadays every gun at his
disposal, and five times that number had he possessed them, might have
opened on any point in the British position he had selected, as soon as
it became light enough to see.

Or, to take a more modern instance, viz., the battle of St.
Privat-Gravelotte, August 18, 1870, where the Germans were able to
concentrate on both wings batteries of two hundred guns and upwards,
it would have been practically impossible, owing to the section of the
slopes of the French position, to carry out the old-fashioned case-shot
attack at all. Nowadays there would be no difficulty in turning on the
fire of two thousand guns on any point of the position, and switching
this fire up and down the line like water from a fire-engine hose, if
the occasion demanded such concentration.

But these alterations in method make no difference in the truth of the
picture of War which Clausewitz presents, with which every soldier, and
above all every Leader, should be saturated.

Death, wounds, suffering, and privation remain the same, whatever the
weapons employed, and their reaction on the ultimate nature of man is
the same now as in the struggle a century ago. It is this reaction that
the Great Commander has to understand and prepare himself to control;
and the task becomes ever greater as, fortunately for humanity, the
opportunities for gathering experience become more rare.

In the end, and with every improvement in science, the result depends
more and more on the character of the Leader and his power of resisting
"the sensuous impressions of the battlefield." Finally, for those who
would fit themselves in advance for such responsibility, I know of no
more inspiring advice than that given by Krishna to Arjuna ages ago,
when the latter trembled before the awful responsibility of launching
his Army against the hosts of the Pandav's:

> This Life within all living things, my Prince,
> Hides beyond harm. Scorn thou to suffer, then,
> For that which cannot suffer. Do thy part!
> Be mindful of thy name, and tremble not.
> Nought better can betide a martial soul
> Than lawful war. Happy the warrior
> To whom comes joy of battle....
> . . . But if thou shunn'st
> This honourable field--a Kshittriya--

If, knowing thy duty and thy task, thou bidd'st
Duty and task go by--that shall be sin!
And those to come shall speak thee infamy
From age to age. But infamy is worse
For men of noble blood to bear than death!

.   .   .   .

Therefore arise, thou Son of Kunti! Brace
Thine arm for conflict; nerve thy heart to meet,
As things alike to thee, pleasure or pain,
Profit or ruin, victory or defeat.
So minded, gird thee to the fight, for so
Thou shalt not sin!

COL. F. N. MAUDE, C.B., late R.E.

CONTENTS

BOOK I ON THE NATURE OF WAR

I   WHAT IS WAR?   page 1
II   END AND MEANS IN WAR   27
III THE GENIUS FOR WAR   46
IV   OF DANGER IN WAR   71
V   OF BODILY EXERTION IN WAR   73
VI   INFORMATION IN WAR   75
VII FRICTION IN WAR   77
VIII   CONCLUDING REMARKS   81

BOOK II ON THE THEORY OF WAR
I   BRANCHES OF THE ART OF WAR 84
II   ON THE THEORY OF WAR 95
III ART OR SCIENCE OF WAR 119
IV   METHODICISM 122V   CRITICISM 130
VI   ON EXAMPLES 156

BOOK III OF STRATEGY IN GENERAL
I   STRATEGY 165
II   ELEMENTS OF STRATEGY   175
III MORAL FORCES   177
IV   THE CHIEF MORAL POWERS   179
V   MILITARY VIRTUE OF AN ARMY   180
VI   BOLDNESS 186
VII PERSEVERANCE   191
VIII   SUPERIORITY OF NUMBERS   192
IX   THE SURPRISE   199
X   STRATAGEM 205
XI   ASSEMBLY OF FORCES IN SPACE   207
XII ASSEMBLY OF FORCES IN TIME   208
XIII   STRATEGIC RESERVE   217
XIV ECONOMY OF FORCES   221
XV   GEOMETRICAL ELEMENT 222
XVI ON THE SUSPENSION OF THE ACT IN WAR   page 224
XVII ON THE CHARACTER OF MODERN WAR 230
XVIII TENSION AND REST 231

BOOK IV THE COMBAT
I   INTRODUCTORY   235
II   CHARACTER OF THE MODERN BATTLE   236
III THE COMBAT IN GENERAL   238
IV   THE COMBAT IN GENERAL (continuation)   243
V   ON THE SIGNIFICATION OF THE COMBAT 253
VI   DURATION OF THE COMBAT   256
VII DECISION OF THE COMBAT   257
VIII   MUTUAL UNDERSTANDING AS TO A BATTLE   266
IX   THE BATTLE   270
X   EFFECTS OF VICTORY 277

On War.txt

XI   THE USE OF THE BATTLE   284
XII STRATEGIC MEANS OF UTILISING VICTORY   292
XIII   RETREAT AFTER A LOST BATTLE   305
XIV NIGHT FIGHTING 308

PREFACE TO THE FIRST EDITION

IT will naturally excite surprise that a preface by a female hand should
accompany a work on such a subject as the present. For my friends no
explanation of the circumstance is required; but I hope by a simple
relation of the cause to clear myself of the appearance of presumption
in the eyes also of those to whom I am not known.

The work to which these lines serve as a preface occupied almost
entirely the last twelve years of the life of my inexpressibly beloved
husband, who has unfortunately been torn too soon from myself and his
country. To complete it was his most earnest desire; but it was not his
intention that it should be published during his life; and if I tried to
persuade him to alter that intention, he often answered, half in jest,
but also, perhaps, half in a foreboding of early death: "Thou shalt
publish it." These words (which in those happy days often drew tears
from me, little as I was inclined to attach a serious meaning to them)
make it now, in the opinion of my friends, a duty incumbent on me
to introduce the posthumous works of my beloved husband, with a few
prefatory lines from myself; and although here may be a difference of
opinion on this point, still I am sure there will be no mistake as to
the feeling which has prompted me to overcome the timidity which makes
any such appearance, even in a subordinate part, so difficult for a
woman.

It will be understood, as a matter of course, that I cannot have the
most remote intention of considering myself as the real editress of a
work which is far above the scope of my capacity: I only stand at its
side as an affectionate companion on its entrance into the world. This
position I may well claim, as a similar one was allowed me during its
formation and progress. Those who are acquainted with our happy married
life, and know how we shared everything with each other--not only
joy and sorrow, but also every occupation, every interest of daily
life--will understand that my beloved husband could not be occupied on
a work of this kind without its being known to me. Therefore, no one can
like me bear testimony to the zeal, to the love with which he laboured
on it, to the hopes which he bound up with it, as well as the manner and
time of its elaboration. His richly gifted mind had from his early youth
longed for light and truth, and, varied as were his talents, still he
had chiefly directed his reflections to the science of war, to which the
duties of his profession called him, and which are of such importance
for the benefit of States. Scharnhorst was the first to lead him into
the right road, and his subsequent appointment in 1810 as Instructor at
the General War School, as well as the honour conferred on him at the
same time of giving military instruction to H.R.H. the Crown Prince,
tended further to give his investigations and studies that direction,
and to lead him to put down in writing whatever conclusions he arrived
at. A paper with which he finished the instruction of H.R.H. the Crown
Prince contains the germ of his subsequent works. But it was in the year
1816, at Coblentz, that he first devoted himself again to scientific
labours, and to collecting the fruits which his rich experience in those
four eventful years had brought to maturity. He wrote down his views,
in the first place, in short essays, only loosely connected with each
other. The following, without date, which has been found amongst his
papers, seems to belong to those early days.

"In the principles here committed to paper, in my opinion, the chief
things which compose Strategy, as it is called, are touched upon. I
looked upon them only as materials, and had just got to such a length
towards the moulding them into a whole.

"These materials have been amassed without any regularly preconceived plan. My view was at first, without regard to system and strict connection, to put down the results of my reflections upon the most important points in quite brief, precise, compact propositions. The manner in which Montesquieu has treated his subject floated before me in idea. I thought that concise, sententious chapters, which I proposed at first to call grains, would attract the attention of the intelligent just as much by that which was to be developed from them, as by that which they contained in themselves. I had, therefore, before me in idea, intelligent readers already acquainted with the subject. But my nature, which always impels me to development and systematising, at last worked its way out also in this instance. For some time I was able to confine myself to extracting only the most important results from the essays, which, to attain clearness and conviction in my own mind, I wrote upon different subjects, to concentrating in that manner their spirit in a small compass; but afterwards my peculiarity gained ascendency completely--I have developed what I could, and thus naturally have supposed a reader not yet acquainted with the subject.

"The more I advanced with the work, and the more I yielded to the spirit of investigation, so much the more I was also led to system; and thus, then, chapter after chapter has been inserted.

"My ultimate view has now been to go through the whole once more, to establish by further explanation much of the earlier treatises, and perhaps to condense into results many analyses on the later ones, and thus to make a moderate whole out of it, forming a small octavo volume. But it was my wish also in this to avoid everything common, everything that is plain of itself, that has been said a hundred times, and is generally accepted; for my ambition was to write a book that would not be forgotten in two or three years, and which any one interested in the subject would at all events take up more than once."

In Coblentz, where he was much occupied with duty, he could only give occasional hours to his private studies. It was not until 1818, after his appointment as Director of the General Academy of War at Berlin, that he had the leisure to expand his work, and enrich it from the history of modern wars. This leisure also reconciled him to his new avocation, which, in other respects, was not satisfactory to him, as, according to the existing organisation of the Academy, the scientific part of the course is not under the Director, but conducted by a Board of Studies. Free as he was from all petty vanity, from every feeling of restless, egotistical ambition, still he felt a desire to be really useful, and not to leave inactive the abilities with which God had endowed him. In active life he was not in a position in which this longing could be satisfied, and he had little hope of attaining to any such position: his whole energies were therefore directed upon the domain of science, and the benefit which he hoped to lay the foundation of by his work was the object of his life. That, notwithstanding this, the resolution not to let the work appear until after his death became more confirmed is the best proof that no vain, paltry longing for praise and distinction, no particle of egotistical views, was mixed up with this noble aspiration for great and lasting usefulness.

Thus he worked diligently on, until, in the spring of 1830, he was appointed to the artillery, and his energies were called into activity in such a different sphere, and to such a high degree, that he was obliged, for the moment at least, to give up all literary work. He then put his papers in order, sealed up the separate packets, labelled them, and took sorrowful leave of this employment which he loved so much. He was sent to Breslau in August of the same year, as Chief of the Second Artillery District, but in December recalled to Berlin, and appointed Chief of the Staff to Field-Marshal Count Gneisenau (for the term of his command). In March 1831, he accompanied his revered Commander to Posen. When he returned from there to Breslau in November after the melancholy event which had taken place, he hoped to resume his work and perhaps complete it in the course of the winter. The Almighty has willed it should be otherwise. On the 7th November he returned to Breslau; on the 16th he was no more; and the packets sealed by himself were not opened until after his death.

The papers thus left are those now made public in the following volumes, exactly in the condition in which they were found, without a word being added or erased. Still, however, there was much to do before publication, in the way of putting them in order and consulting about them; and I am deeply indebted to several sincere friends for the assistance they have afforded me, particularly Major O'Etzel, who kindly undertook the correction of the Press, as well as the preparation of the maps to accompany the historical parts of the work. I must also mention my much-loved brother, who was my support in the hour of my misfortune, and who has also done much for me in respect of these papers; amongst other things, by carefully examining and putting them in order, he found the commencement of the revision which my dear husband wrote in the year 1827, and mentions in the Notice hereafter annexed as a work he had in view. This revision has been inserted in the place intended for it in the first book (for it does not go any further).

There are still many other friends to whom I might offer my thanks for their advice, for the sympathy and friendship which they have shown me; but if I do not name them all, they will, I am sure, not have any doubts of my sincere gratitude. It is all the greater, from my firm conviction that all they have done was not only on my own account, but for the friend whom God has thus called away from them so soon.

If I have been highly blessed as the wife of such a man during one and twenty years, so am I still, notwithstanding my irreparable loss, by the treasure of my recollections and of my hopes, by the rich legacy of sympathy and friendship which I owe the beloved departed, by the elevating feeling which I experience at seeing his rare worth so generally and honourably acknowledged.

The trust confided to me by a Royal Couple is a fresh benefit for which I have to thank the Almighty, as it opens to me an honourable occupation, to which Idevote myself. May this occupation be blessed, and may the dear little Prince who is now entrusted to my care, some day read this book, and be animated by it to deeds like those of his glorious ancestors.

Written at the Marble Palace, Potsdam, 30th June, 1832.

MARIE VON CLAUSEWITZ, Born Countess Bruhl, Oberhofmeisterinn to H.R.H. the Princess William.

NOTICE

I LOOK upon the first six books, of which a fair copy has now been made, as only a mass which is still in a manner without form, and which has yet to be again revised. In this revision the two kinds of War will be everywhere kept more distinctly in view, by which all ideas will acquire a clearer meaning, a more precise direction, and a closer application. The two kinds of War are, first, those in which the object is the OVERTHROW OF THE ENEMY, whether it be that we aim at his destruction, politically, or merely at disarming him and forcing him to conclude peace on our terms; and next, those in which our object is MERELY TO MAKE SOME CONQUESTS ON THE FRONTIERS OF HIS COUNTRY, either for the purpose of retaining them permanently, or of turning them to account as matter of exchange in the settlement of a peace. Transition from one kind to the other must certainly continue to exist, but the completely different nature of the tendencies of the two must everywhere appear, and must separate from each other things which are incompatible.

Besides establishing this real difference in Wars, another practically
necessary point of view must at the same time be established, which is,
that WAR IS ONLY A CONTINUATION OF STATE POLICY BY OTHER MEANS. This
point of view being adhered to everywhere, will introduce much more
unity into the consideration of the subject, and things will be more
easily disentangled from each other. Although the chief application of
this point of view does not commence until we get to the eighth book,
still it must be completely developed in the first book, and also lend
assistance throughout the revision of the first six books. Through such
a revision the first six books will get rid of a good deal of dross,
many rents and chasms will be closed up, and much that is of a general
nature will be transformed into distinct conceptions and forms.

The seventh book--on attack--for the different chapters of which
sketches are already made, is to be considered as a reflection of the
sixth, and must be completed at once, according to the above-mentioned
more distinct points of view, so that it will require no fresh revision,
but rather may serve as a model in the revision of the first six books.

For the eighth book--on the Plan of a War, that is, of the organisation
of a whole War in general--several chapters are designed, but they are
not at all to be regarded as real materials, they are merely a track,
roughly cleared, as it were, through the mass, in order by that means to
ascertain the points of most importance. They have answered this object,
and I propose, on finishing the seventh book, to proceed at once to the
working out of the eighth, where the two points of view above mentioned
will be chiefly affirmed, by which everything will be simplified, and
at the same time have a spirit breathed into it. I hope in this book to
iron out many creases in the heads of strategists and statesmen, and at
least to show the object of action, and the real point to be considered
in War.

Now, when I have brought my ideas clearly out by finishing this eighth
book, and have properly established the leading features of War, it will
be easier for me to carry the spirit of these ideas in to the first
six books, and to make these same features show themselves everywhere.
Therefore I shall defer till then the revision of the first six books.

Should the work be interrupted by my death, then what is found can only
be called a mass of conceptions not brought into form; but as these
are open to endless misconceptions, they will doubtless give rise to a
number of crude criticisms: for in these things, every one thinks, when
he takes up his pen, that whatever comes into his head is worth saying
and printing, and quite as incontrovertible as that twice two make four.
If such a one would take the pains, as I have done, to think over the
subject, for years, and to compare his ideas with military history, he
would certainly be a little more guarded in his criticism.

Still, notwithstanding this imperfect form, I believe that an impartial
reader thirsting for truth and conviction will rightly appreciate in the
first six books the fruits of several years' reflection and a diligent
study of War, and that, perhaps, he will find in them some leading ideas
which may bring about a revolution in the theory of War.

Berlin, 10th July, 1827.

Besides this notice, amongst the papers left the following unfinished
memorandum was found, which appears of very recent date:

The manuscript on the conduct of the Grande Guerre, which will be
found after my death, in its present state can only be regarded as a
collection of materials from which it is intended to construct a theory
of War. With the greater part I am not yet satisfied; and the sixth book
is to be looked at as a mere essay: I should have completely remodelled
it, and have tried a different line.

But the ruling principles which pervade these materials I hold to be the right ones: they are the result of a very varied reflection, keeping always in view the reality, and always bearing in mind what I have learnt by experience and by my intercourse with distinguished soldiers.

The seventh book is to contain the attack, the subjects of which are thrown together in a hasty manner: the eighth, the plan for a War, in which I would have examined War more especially in its political and human aspects.

The first chapter of the first book is the only one which I consider as completed; it will at least serve to show the manner in which I proposed to treat the subject throughout.

The theory of the Grande Guerre, or Strategy, as it is called, is beset with extraordinary difficulties, and we may affirm that very few men have clear conceptions of the separate subjects, that is, conceptions carried up to their full logical conclusions. In real action most men are guided merely by the tact of judgment which hits the object more or less accurately, according as they possess more or less genius.

This is the way in which all great Generals have acted, and therein partly lay their greatness and their genius, that they always hit upon what was right by this tact. Thus also it will always be in action, and so far this tact is amply sufficient. But when it is a question, not of acting oneself, but of convincing others in a consultation, then all depends on clear conceptions and demonstration of the inherent relations, and so little progress has been made in this respect that most deliberations are merely a contention of words, resting on no firm basis, and ending either in every one retaining his own opinion, or in a compromise from mutual considerations of respect, a middle course really without any value.(*)

> (*) Herr Clausewitz evidently had before his mind the endless consultations at the Headquarters of the Bohemian Army in the Leipsic Campaign 1813.

Clear ideas on these matters are therefore not wholly useless; besides, the human mind has a general tendency to clearness, and always wants to be consistent with the necessary order of things.

Owing to the great difficulties attending a philosophical construction of the Art of War, and the many attempts at it that have failed, most people have come to the conclusion that such a theory is impossible, because it concerns things which no standing law can embrace. We should also join in this opinion and give up any attempt at a theory, were it not that a great number of propositions make themselves evident without any difficulty, as, for instance, that the defensive form, with a negative object, is the stronger form, the attack, with the positive object, the weaker--that great results carry the little ones with them--that, therefore, strategic effects may be referred to certain centres of gravity--that a demonstration is a weaker application of force than a real attack, that, therefore, there must be some special reason for resorting to the former--that victory consists not merely in the conquest on the field of battle, but in the destruction of armed forces, physically and morally, which can in general only be effected by a pursuit after the battle is gained--that successes are always greatest at the point where the victory has been gained, that, therefore, the change from one line and object to another can only be regarded as a necessary evil--that a turning movement is only justified by a superiority of numbers generally or by the advantage of our lines of communication and retreat over those of the enemy--that flank positions are only justifiable on similar grounds--that every attack becomes weaker as it progresses.

THE INTRODUCTION OF THE AUTHOR

THAT the conception of the scientific does not consist alone, or chiefly, in system, and its finished theoretical constructions, requires nowadays no exposition. System in this treatise is not to be found on the surface, and instead of a finished building of theory, there are only materials.

The scientific form lies here in the endeavour to explore the nature of military phenomena to show their affinity with the nature of the things of which they are composed. Nowhere has the philosophical argument been evaded, but where it runs out into too thin a thread the Author has preferred to cut it short, and fall back upon the corresponding results of experience; for in the same way as many plants only bear fruit when they do not shoot too high, so in the practical arts the theoretical leaves and flowers must not be made to sprout too far, but kept near to experience, which is their proper soil.

Unquestionably it would be a mistake to try to discover from the chemical ingredients of a grain of corn the form of the ear of corn which it bears, as we have only to go to the field to see the ears ripe. Investigation and observation, philosophy and experience, must neither despise nor exclude one another; they mutually afford each other the rights of citizenship. Consequently, the propositions of this book, with their arch of inherent necessity, are supported either by experience or by the conception of War itself as external points, so that they are not without abutments.(*)

> (*) That this is not the case in the works of many military writers especially of those who have aimed at treating of War itself in a scientific manner, is shown in many instances, in which by their reasoning, the pro and contra swallow each other up so effectually that there is no vestige of the tails even which were left in the case of the two lions.

It is, perhaps, not impossible to write a systematic theory of War full of spirit and substance, but ours hitherto, have been very much the reverse. To say nothing of their unscientific spirit, in their striving after coherence and completeness of system, they overflow with commonplaces, truisms, and twaddle of every kind. If we want a striking picture of them we have only to read Lichtenberg's extract from a code of regulations in case of fire.

If a house takes fire, we must seek, above all things, to protect the right side of the house standing on the left, and, on the other hand, the left side of the house on the right; for if we, for example, should protect the left side of the house on the left, then the right side of the house lies to the right of the left, and consequently as the fire lies to the right of this side, and of the right side (for we have assumed that the house is situated to the left of the fire), therefore the right side is situated nearer to the fire than the left, and the right side of the house might catch fire if it was not protected before it came to the left, which is protected. Consequently, something might be burnt that is not protected, and that sooner than something else would be burnt, even if it was not protected; consequently we must let alone the latter and protect the former. In order to impress the thing on one's mind, we have only to note if the house is situated to the right of the fire, then it is the left side, and if the house is to the left it is the right side.

In order not to frighten the intelligent reader by such commonplaces, and to make the little good that there is distasteful by pouring water upon it, the Author has preferred to give in small ingots of fine metal his impressions and convictions, the result of many years' reflection on War, of his intercourse with men of ability, and of much personal experience. Thus the seemingly weakly bound-together chapters of this book have arisen, but it is hoped they will not be found wanting in logical connection. Perhaps soon a greater head may appear, and instead

of these single grains, give the whole in a casting of pure metal
without dross.

BRIEF MEMOIR OF GENERAL CLAUSEWITZ

(BY TRANSLATOR)

THE Author of the work here translated, General Carl Von Clausewitz, was
born at Burg, near Magdeburg, in 1780, and entered the Prussian Army
as Fahnenjunker (i.e., ensign) in 1792. He served in the campaigns of
1793-94 on the Rhine, after which he seems to have devoted some time
to the study of the scientific branches of his profession. In 1801 he
entered the Military School at Berlin, and remained there till
1803. During his residence there he attracted the notice of General
Scharnhorst, then at the head of the establishment; and the patronage of
this distinguished officer had immense influence on his future career,
and we may gather from his writings that he ever afterwards continued
to entertain a high esteem for Scharnhorst. In the campaign of 1806 he
served as Aide-de-camp to Prince Augustus of Prussia; and being wounded
and taken prisoner, he was sent into France until the close of that
war. On his return, he was placed on General Scharnhorst's Staff, and
employed in the work then going on for the reorganisation of the Army.
He was also at this time selected as military instructor to the late
King of Prussia, then Crown Prince. In 1812 Clausewitz, with several
other Prussian officers, having entered the Russian service, his first
appointment was as Aide-de-camp to General Phul. Afterwards, while
serving with Wittgenstein's army, he assisted in negotiating the famous
convention of Tauroggen with York. Of the part he took in that affair he
has left an interesting account in his work on the "Russian Campaign."
It is there stated that, in order to bring the correspondence which had
been carried on with York to a termination in one way or another, the
Author was despatched to York's headquarters with two letters, one was
from General d'Auvray, the Chief of the Staff of Wittgenstein's army, to
General Diebitsch, showing the arrangements made to cut off York's corps
from Macdonald (this was necessary in order to give York a plausible
excuse for seceding from the French); the other was an intercepted
letter from Macdonald to the Duke of Bassano. With regard to the former
of these, the Author says, "it would not have had weight with a man like
York, but for a military justification, if the Prussian Court should
require one as against the French, it was important."

The second letter was calculated at the least to call up in General
York's mind all the feelings of bitterness which perhaps for some days
past bad been diminished by the consciousness of his own behaviour
towards the writer.

As the Author entered General York's chamber, the latter called out to
him, "Keep off from me; I will have nothing more to do with you; your
d----d Cossacks have let a letter of Macdonald's pass through them,
which brings me an order to march on Piktrepohnen, in order there to
effect our junction. All doubt is now at an end; your troops do not
come up; you are too weak; march I must, and I must excuse myself from
further negotiation, which may cost me my head." The Author said that be
would make no opposition to all this, but begged for a candle, as he
had letters to show the General, and, as the latter seemed still to
hesitate, the Author added, "Your Excellency will not surely place me in
the embarrassment of departing without having executed my commission."
The General ordered candles, and called in Colonel von Roeder, the chief
of his staff, from the ante-chamber. The letters were read. After a
pause of an instant, the General said, "Clausewitz, you are a Prussian,
do you believe that the letter of General d'Auvray is sincere, and that
Wittgenstein's troops will really be at the points he mentioned on the
31st?" The Author replied, "I pledge myself for the sincerity of this
letter upon the knowledge I have of General d'Auvray and the other men
of Wittgenstein's headquarters; whether the dispositions he announces

On War.txt

can be accomplished as he lays down I certainly cannot pledge myself;
for your Excellency knows that in war we must often fall short of the
line we have drawn for ourselves." The General was silent for a few
minutes of earnest reflection; then he held out his hand to the Author,
and said, "You have me. Tell General Diebitsch that we must confer early
to-morrow at the mill of Poschenen, and that I am now firmly determined
to separate myself from the French and their cause." The hour was fixed
for 8 A.M. After this was settled, the General added, "But I will not
do the thing by halves, I will get you Massenbach also." He called in
an officer who was of Massenbach's cavalry, and who had just left them.
Much like Schiller's Wallenstein, he asked, walking up and down the
room the while, "What say your regiments?" The officer broke out with
enthusiasm at the idea of a riddance from the French alliance, and said
that every man of the troops in question felt the same.

"You young ones may talk; but my older head is shaking on my shoulders,"
replied the General.(*)

(*) "Campaign in Russia in 1812"; translated from the German
of General Von Clausewitz (by Lord Ellesmere).

After the close of the Russian campaign Clausewitz remained in the
service of that country, but was attached as a Russian staff officer to
Blucher's headquarters till the Armistice in 1813.

In 1814, he became Chief of the Staff of General Walmoden's Russo-German
Corps, which formed part of the Army of the North under Bernadotte.
His name is frequently mentioned with distinction in that campaign,
particularly in connection with the affair of Goehrde.

Clausewitz re-entered the Prussian service in 1815, and served as Chief
of the Staff to Thielman's corps, which was engaged with Grouchy at
Wavre, on the 18th of June.

After the Peace, he was employed in a command on the Rhine. In 1818, he
became Major-General, and Director of the Military School at which he
had been previously educated.

In 1830, he was appointed Inspector of Artillery at Breslau, but soon
after nominated Chief of the Staff to the Army of Observation, under
Marshal Gneisenau on the Polish frontier.

The latest notices of his life and services are probably to be found
in the memoirs of General Brandt, who, from being on the staff of
Gneisenau's army, was brought into daily intercourse with Clausewitz
in matters of duty, and also frequently met him at the table of Marshal
Gneisenau, at Posen.

Amongst other anecdotes, General Brandt relates that, upon one occasion,
the conversation at the Marshal's table turned upon a sermon preached
by a priest, in which some great absurdities were introduced, and a
discussion arose as to whether the Bishop should not be made responsible
for what the priest had said. This led to the topic of theology in
general, when General Brandt, speaking of himself, says, "I expressed an
opinion that theology is only to be regarded as an historical process,
as a MOMENT in the gradual development of the human race. This
brought upon me an attack from all quarters, but more especially
from Clausewitz, who ought to have been on my side, he having been an
adherent and pupil of Kiesewetter's, who had indoctrinated him in the
philosophy of Kant, certainly diluted--I might even say in homoeopathic
doses." This anecdote is only interesting as the mention of Kiesewetter
points to a circumstance in the life of Clausewitz that may have had
an influence in forming those habits of thought which distinguish his
writings.

"The way," says General Brandt, "in which General Clausewitz judged
of things, drew conclusions from movements and marches, calculated the
times of the marches, and the points where decisions would take

place, was extremely interesting. Fate has unfortunately denied him an opportunity of showing his talents in high command, but I have a firm persuasion that as a strategist he would have greatly distinguished himself. As a leader on the field of battle, on the other hand, he would not have been so much in his right place, from a manque d'habitude du commandement, he wanted the art d'enlever les troupes."

After the Prussian Army of Observation was dissolved, Clausewitz returned to Breslau, and a few days after his arrival was seized with cholera, the seeds of which he must have brought with him from the army on the Polish frontier. His death took place in November 1831.

His writings are contained in nine volumes, published after his death, but his fame rests most upon the three volumes forming his treatise on "War." In the present attempt to render into English this portion of the works of Clausewitz, the translator is sensible of many deficiencies, but he hopes at all events to succeed in making this celebrated treatise better known in England, believing, as he does, that so far as the work concerns the interests of this country, it has lost none of the importance it possessed at the time of its first publication.

J. J. GRAHAM (Col.)

BOOK I. ON THE NATURE OF WAR

CHAPTER I. WHAT IS WAR?

1. INTRODUCTION.

WE propose to consider first the single elements of our subject, then each branch or part, and, last of all, the whole, in all its relations--therefore to advance from the simple to the complex. But it is necessary for us to commence with a glance at the nature of the whole, because it is particularly necessary that in the consideration of any of the parts their relation to the whole should be kept constantly in view.

2. DEFINITION.

We shall not enter into any of the abstruse definitions of War used by publicists. We shall keep to the element of the thing itself, to a duel. War is nothing but a duel on an extensive scale. If we would conceive as a unit the countless number of duels which make up a War, we shall do so best by supposing to ourselves two wrestlers. Each strives by physical force to compel the other to submit to his will: each endeavours to throw his adversary, and thus render him incapable of further resistance.

WAR THEREFORE IS AN ACT OF VIOLENCE INTENDED TO COMPEL OUR OPPONENT TO FULFIL OUR WILL.

Violence arms itself with the inventions of Art and Science in order to contend against violence. Self-imposed restrictions, almost imperceptible and hardly worth mentioning, termed usages of International Law, accompany it without essentially impairing its power. Violence, that is to say, physical force (for there is no moral force without the conception of States and Law), is therefore the MEANS; the compulsory submission of the enemy to our will is the ultimate object. In order to attain this object fully, the enemy must be disarmed, and disarmament becomes therefore the immediate OBJECT of hostilities in theory. It takes the place of the final object, and puts it aside as something we can eliminate from our calculations.

3. UTMOST USE OF FORCE.

Now, philanthropists may easily imagine there is a skilful method of disarming and overcoming an enemy without great bloodshed, and that this is the proper tendency of the Art of War. However plausible this may appear, still it is an error which must be extirpated; for in such dangerous things as War, the errors which proceed from a spirit of benevolence are the worst. As the use of physical power to the utmost extent by no means excludes the co-operation of the intelligence, it follows that he who uses force unsparingly, without reference to the bloodshed involved, must obtain a superiority if his adversary uses less vigour in its application. The former then dictates the law to the latter, and both proceed to extremities to which the only limitations are those imposed by the amount of counter-acting force on each side.

This is the way in which the matter must be viewed and it is to no purpose, it is even against one's own interest, to turn away from the consideration of the real nature of the affair because the horror of its elements excites repugnance.

If the Wars of civilised people are less cruel and destructive than those of savages, the difference arises from the social condition both of States in themselves and in their relations to each other. Out of this social condition and its relations War arises, and by it War is subjected to conditions, is controlled and modified. But these things do not belong to War itself; they are only given conditions; and to introduce into the philosophy of War itself a principle of moderation would be an absurdity.

Two motives lead men to War: instinctive hostility and hostile intention. In our definition of War, we have chosen as its characteristic the latter of these elements, because it is the most general. It is impossible to conceive the passion of hatred of the wildest description, bordering on mere instinct, without combining with it the idea of a hostile intention. On the other hand, hostile intentions may often exist without being accompanied by any, or at all events by any extreme, hostility of feeling. Amongst savages views emanating from the feelings, amongst civilised nations those emanating from the understanding, have the predominance; but this difference arises from attendant circumstances, existing institutions, &c., and, therefore, is not to be found necessarily in all cases, although it prevails in the majority. In short, even the most civilised nations may burn with passionate hatred of each other.

We may see from this what a fallacy it would be to refer the War of a civilised nation entirely to an intelligent act on the part of the Government, and to imagine it as continually freeing itself more and more from all feeling of passion in such a way that at last the physical masses of combatants would no longer be required; in reality, their mere relations would suffice--a kind of algebraic action.

Theory was beginning to drift in this direction until the facts of the last War(*) taught it better. If War is an ACT of force, it belongs necessarily also to the feelings. If it does not originate in the feelings, it REACTS, more or less, upon them, and the extent of this reaction depends not on the degree of civilisation, but upon the importance and duration of the interests involved.

   (*) Clausewitz alludes here to the "Wars of Liberation,"
   1813,14,15.

Therefore, if we find civilised nations do not put their prisoners to death, do not devastate towns and countries, this is because their intelligence exercises greater influence on their mode of carrying on War, and has taught them more effectual means of applying force than these rude acts of mere instinct. The invention of gunpowder, the constant progress of improvements in the construction of firearms, are sufficient proofs that the tendency to destroy the adversary which lies at the bottom of the conception of War is in no way changed or modified through the progress of civilisation.

We therefore repeat our proposition, that War is an act of violence
pushed to its utmost bounds; as one side dictates the law to the other,
there arises a sort of reciprocal action, which logically must lead to
an extreme. This is the first reciprocal action, and the first extreme
with which we meet (FIRST RECIPROCAL ACTION).

4. THE AIM IS TO DISARM THE ENEMY.

We have already said that the aim of all action in War is to disarm
the enemy, and we shall now show that this, theoretically at least, is
indispensable.

If our opponent is to be made to comply with our will, we must place him
in a situation which is more oppressive to him than the sacrifice which
we demand; but the disadvantages of this position must naturally not
be of a transitory nature, at least in appearance, otherwise the enemy,
instead of yielding, will hold out, in the prospect of a change for
the better. Every change in this position which is produced by a
continuation of the War should therefore be a change for the worse. The
worst condition in which a belligerent can be placed is that of being
completely disarmed. If, therefore, the enemy is to be reduced to
submission by an act of War, he must either be positively disarmed or
placed in such a position that he is threatened with it. From this it
follows that the disarming or overthrow of the enemy, whichever we call
it, must always be the aim of Warfare. Now War is always the shock of
two hostile bodies in collision, not the action of a living power upon
an inanimate mass, because an absolute state of endurance would not be
making War; therefore, what we have just said as to the aim of action in
War applies to both parties. Here, then, is another case of reciprocal
action. As long as the enemy is not defeated, he may defeat me; then I
shall be no longer my own master; he will dictate the law to me as I
did to him. This is the second reciprocal action, and leads to a second
extreme (SECOND RECIPROCAL ACTION).

5. UTMOST EXERTION OF POWERS.

If we desire to defeat the enemy, we must proportion our efforts to his
powers of resistance. This is expressed by the product of two factors
which cannot be separated, namely, the sum of available means and the
strength of the Will. The sum of the available means may be estimated in
a measure, as it depends (although not entirely) upon numbers; but the
strength of volition is more difficult to determine, and can only be
estimated to a certain extent by the strength of the motives. Granted we
have obtained in this way an approximation to the strength of the power
to be contended with, we can then take of our own means, and either
increase them so as to obtain a preponderance, or, in case we have not
the resources to effect this, then do our best by increasing our means
as far as possible. But the adversary does the same; therefore, there is
a new mutual enhancement, which, in pure conception, must create a fresh
effort towards an extreme. This is the third case of reciprocal action,
and a third extreme with which we meet (THIRD RECIPROCAL ACTION).

6. MODIFICATION IN THE REALITY.

Thus reasoning in the abstract, the mind cannot stop short of an
extreme, because it has to deal with an extreme, with a conflict of
forces left to themselves, and obeying no other but their own inner
laws. If we should seek to deduce from the pure conception of War an
absolute point for the aim which we shall propose and for the means
which we shall apply, this constant reciprocal action would involve us
in extremes, which would be nothing but a play of ideas produced by an
almost invisible train of logical subtleties. If, adhering closely to
the absolute, we try to avoid all difficulties by a stroke of the pen,
and insist with logical strictness that in every case the extreme must
be the object, and the utmost effort must be exerted in that direction,
such a stroke of the pen would be a mere paper law, not by any means
adapted to the real world.

Even supposing this extreme tension of forces was an absolute which
could easily be ascertained, still we must admit that the human mind
would hardly submit itself to this kind of logical chimera. There
would be in many cases an unnecessary waste of power, which would be in
opposition to other principles of statecraft; an effort of will would
be required disproportioned to the proposed object, which therefore it
would be impossible to realise, for the human will does not derive its
impulse from logical subtleties.

But everything takes a different shape when we pass from abstractions to
reality. In the former, everything must be subject to optimism, and we
must imagine the one side as well as the other striving after perfection
and even attaining it. Will this ever take place in reality? It will if,

(1) War becomes a completely isolated act, which arises suddenly, and is
in no way connected with the previous history of the combatant States.

(2) If it is limited to a single solution, or to several simultaneous
solutions.

(3) If it contains within itself the solution perfect and complete,
free from any reaction upon it, through a calculation beforehand of the
political situation which will follow from it.

7. WAR IS NEVER AN ISOLATED ACT.

With regard to the first point, neither of the two opponents is an
abstract person to the other, not even as regards that factor in the sum
of resistance which does not depend on objective things, viz., the will.
This Will is not an entirely unknown quantity; it indicates what it
will be to-morrow by what it is to-day. War does not spring up quite
suddenly, it does not spread to the full in a moment; each of the two
opponents can, therefore, form an opinion of the other, in a great
measure, from what he is and what he does, instead of judging of him
according to what he, strictly speaking, should be or should do. But,
now, man with his incomplete organisation is always below the line of
absolute perfection, and thus these deficiencies, having an influence on
both sides, become a modifying principle.

8. WAR DOES NOT CONSIST OF A SINGLE INSTANTANEOUS BLOW.

The second point gives rise to the following considerations:--

If war ended in a single solution, or a number of simultaneous ones,
then naturally all the preparations for the same would have a tendency
to the extreme, for an omission could not in any way be repaired; the
utmost, then, that the world of reality could furnish as a guide for us
would be the preparations of the enemy, as far as they are known to
us; all the rest would fall into the domain of the abstract. But if
the result is made up from several successive acts, then naturally that
which precedes with all its phases may be taken as a measure for that
which will follow, and in this manner the world of reality again takes
the place of the abstract, and thus modifies the effort towards the
extreme.

Yet every War would necessarily resolve itself into a single solution,
or a sum of simultaneous results, if all the means required for the
struggle were raised at once, or could be at once raised; for as one
adverse result necessarily diminishes the means, then if all the means
have been applied in the first, a second cannot properly be supposed.
All hostile acts which might follow would belong essentially to the
first, and form, in reality only its duration.

But we have already seen that even in the preparation for War the real
world steps into the place of mere abstract conception--a material
standard into the place of the hypotheses of an extreme: that therefore
Page 17

in that way both parties, by the influence of the mutual reaction, remain below the line of extreme effort, and therefore all forces are not at once brought forward.

It lies also in the nature of these forces and their application that they cannot all be brought into activity at the same time. These forces are THE ARMIES ACTUALLY ON FOOT, THE COUNTRY, with its superficial extent and its population, AND THE ALLIES.

In point of fact, the country, with its superficial area and the population, besides being the source of all military force, constitutes in itself an integral part of the efficient quantities in War, providing either the theatre of war or exercising a considerable influence on the same.

Now, it is possible to bring all the movable military forces of a country into operation at once, but not all fortresses, rivers, mountains, people, &c.--in short, not the whole country, unless it is so small that it may be completely embraced by the first act of the War. Further, the co-operation of allies does not depend on the Will of the belligerents; and from the nature of the political relations of states to each other, this co-operation is frequently not afforded until after the War has commenced, or it may be increased to restore the balance of power.

That this part of the means of resistance, which cannot at once be brought into activity, in many cases, is a much greater part of the whole than might at first be supposed, and that it often restores the balance of power, seriously affected by the great force of the first decision, will be more fully shown hereafter. Here it is sufficient to show that a complete concentration of all available means in a moment of time is contradictory to the nature of War.

Now this, in itself, furnishes no ground for relaxing our efforts to accumulate strength to gain the first result, because an unfavourable issue is always a disadvantage to which no one would purposely expose himself, and also because the first decision, although not the only one, still will have the more influence on subsequent events, the greater it is in itself.

But the possibility of gaining a later result causes men to take refuge in that expectation, owing to the repugnance in the human mind to making excessive efforts; and therefore forces are not concentrated and measures are not taken for the first decision with that energy which would otherwise be used. Whatever one belligerent omits from weakness, becomes to the other a real objective ground for limiting his own efforts, and thus again, through this reciprocal action, extreme tendencies are brought down to efforts on a limited scale.

9. THE RESULT IN WAR IS NEVER ABSOLUTE.

Lastly, even the final decision of a whole War is not always to be regarded as absolute. The conquered State often sees in it only a passing evil, which may be repaired in after times by means of political combinations. How much this must modify the degree of tension, and the vigour of the efforts made, is evident in itself.

10. THE PROBABILITIES OF REAL LIFE TAKE THE PLACE OF THE CONCEPTIONS OF THE EXTREME AND THE ABSOLUTE.

In this manner, the whole act of War is removed from the rigorous law of forces exerted to the utmost. If the extreme is no longer to be apprehended, and no longer to be sought for, it is left to the judgment to determine the limits for the efforts to be made in place of it, and this can only be done on the data furnished by the facts of the real world by the LAWS OF PROBABILITY. Once the belligerents are no longer

mere conceptions, but individual States and Governments, once the War
is no longer an ideal, but a definite substantial procedure, then the
reality will furnish the data to compute the unknown quantities which
are required to be found.

From the character, the measures, the situation of the adversary,
and the relations with which he is surrounded, each side will draw
conclusions by the law of probability as to the designs of the other,
and act accordingly.

11. THE POLITICAL OBJECT NOW REAPPEARS.

Here the question which we had laid aside forces itself again into
consideration (see No. 2), viz., the political object of the War. The
law of the extreme, the view to disarm the adversary, to overthrow
him, has hitherto to a certain extent usurped the place of this end or
object. Just as this law loses its force, the political must again come
forward. If the whole consideration is a calculation of probability
based on definite persons and relations, then the political object,
being the original motive, must be an essential factor in the product.
The smaller the sacrifice we demand from ours, the smaller, it may be
expected, will be the means of resistance which he will employ; but the
smaller his preparation, the smaller will ours require to be. Further,
the smaller our political object, the less value shall we set upon it,
and the more easily shall we be induced to give it up altogether.

Thus, therefore, the political object, as the original motive of the
War, will be the standard for determining both the aim of the military
force and also the amount of effort to be made. This it cannot be in
itself, but it is so in relation to both the belligerent States, because
we are concerned with realities, not with mere abstractions. One and
the same political object may produce totally different effects upon
different people, or even upon the same people at different times;
we can, therefore, only admit the political object as the measure, by
considering it in its effects upon those masses which it is to move, and
consequently the nature of those masses also comes into consideration.
It is easy to see that thus the result may be very different according
as these masses are animated with a spirit which will infuse vigour
into the action or otherwise. It is quite possible for such a state
of feeling to exist between two States that a very trifling political
motive for War may produce an effect quite disproportionate--in fact, a
perfect explosion.

This applies to the efforts which the political object will call forth
in the two States, and to the aim which the military action shall
prescribe for itself. At times it may itself be that aim, as, for
example, the conquest of a province. At other times the political object
itself is not suitable for the aim of military action; then such a one
must be chosen as will be an equivalent for it, and stand in its place
as regards the conclusion of peace. But also, in this, due attention to
the peculiar character of the States concerned is always supposed. There
are circumstances in which the equivalent must be much greater than the
political object, in order to secure the latter. The political object
will be so much the more the standard of aim and effort, and have more
influence in itself, the more the masses are indifferent, the less that
any mutual feeling of hostility prevails in the two States from other
causes, and therefore there are cases where the political object almost
alone will be decisive.

If the aim of the military action is an equivalent for the political
object, that action will in general diminish as the political object
diminishes, and in a greater degree the more the political object
dominates. Thus it is explained how, without any contradiction in
itself, there may be Wars of all degrees of importance and energy, from
a War of extermination down to the mere use of an army of observation.
This, however, leads to a question of another kind which we have
hereafter to develop and answer.

12. A SUSPENSION IN THE ACTION OF WAR UNEXPLAINED BY ANYTHING SAID AS YET.

However insignificant the political claims mutually advanced, however weak the means put forth, however small the aim to which military action is directed, can this action be suspended even for a moment? This is a question which penetrates deeply into the nature of the subject.

Every transaction requires for its accomplishment a certain time which we call its duration. This may be longer or shorter, according as the person acting throws more or less despatch into his movements.

About this more or less we shall not trouble ourselves here. Each person acts in his own fashion; but the slow person does not protract the thing because he wishes to spend more time about it, but because by his nature he requires more time, and if he made more haste would not do the thing so well. This time, therefore, depends on subjective causes, and belongs to the length, so called, of the action.

If we allow now to every action in War this, its length, then we must assume, at first sight at least, that any expenditure of time beyond this length, that is, every suspension of hostile action, appears an absurdity; with respect to this it must not be forgotten that we now speak not of the progress of one or other of the two opponents, but of the general progress of the whole action of the War.

13. THERE IS ONLY ONE CAUSE WHICH CAN SUSPEND THE ACTION, AND THIS SEEMS TO BE ONLY POSSIBLE ON ONE SIDE IN ANY CASE.

If two parties have armed themselves for strife, then a feeling of animosity must have moved them to it; as long now as they continue armed, that is, do not come to terms of peace, this feeling must exist; and it can only be brought to a standstill by either side by one single motive alone, which is, THAT HE WAITS FOR A MORE FAVOURABLE MOMENT FOR ACTION. Now, at first sight, it appears that this motive can never exist except on one side, because it, eo ipso, must be prejudicial to the other. If the one has an interest in acting, then the other must have an interest in waiting.

A complete equilibrium of forces can never produce a suspension of action, for during this suspension he who has the positive object (that is, the assailant) must continue progressing; for if we should imagine an equilibrium in this way, that he who has the positive object, therefore the strongest motive, can at the same time only command the lesser means, so that the equation is made up by the product of the motive and the power, then we must say, if no alteration in this condition of equilibrium is to be expected, the two parties must make peace; but if an alteration is to be expected, then it can only be favourable to one side, and therefore the other has a manifest interest to act without delay. We see that the conception of an equilibrium cannot explain a suspension of arms, but that it ends in the question of the EXPECTATION OF A MORE FAVOURABLE MOMENT.

Let us suppose, therefore, that one of two States has a positive object, as, for instance, the conquest of one of the enemy's provinces--which is to be utilised in the settlement of peace. After this conquest, his political object is accomplished, the necessity for action ceases, and for him a pause ensues. If the adversary is also contented with this solution, he will make peace; if not, he must act. Now, if we suppose that in four weeks he will be in a better condition to act, then he has sufficient grounds for putting off the time of action.

But from that moment the logical course for the enemy appears to be to act that he may not give the conquered party THE DESIRED time. Of course, in this mode of reasoning a complete insight into the state of circumstances on both sides is supposed.

14. THUS A CONTINUANCE OF ACTION WILL ENSUE WHICH WILL ADVANCE TOWARDS A CLIMAX.

If this unbroken continuity of hostile operations really existed, the effect would be that everything would again be driven towards the extreme; for, irrespective of the effect of such incessant activity in inflaming the feelings, and infusing into the whole a greater degree of passion, a greater elementary force, there would also follow from this continuance of action a stricter continuity, a closer connection between cause and effect, and thus every single action would become of more importance, and consequently more replete with danger.

But we know that the course of action in War has seldom or never this unbroken continuity, and that there have been many Wars in which action occupied by far the smallest portion of time employed, the whole of the rest being consumed in inaction. It is impossible that this should be always an anomaly; suspension of action in War must therefore be possible, that is no contradiction in itself. We now proceed to show how this is.

15. HERE, THEREFORE, THE PRINCIPLE OF POLARITY IS BROUGHT INTO REQUISITION.

As we have supposed the interests of one Commander to be always antagonistic to those of the other, we have assumed a true POLARITY. We reserve a fuller explanation of this for another chapter, merely making the following observation on it at present.

The principle of polarity is only valid when it can be conceived in one and the same thing, where the positive and its opposite the negative completely destroy each other. In a battle both sides strive to conquer; that is true polarity, for the victory of the one side destroys that of the other. But when we speak of two different things which have a common relation external to themselves, then it is not the things but their relations which have the polarity.

16. ATTACK AND DEFENCE ARE THINGS DIFFERING IN KIND AND OF UNEQUAL FORCE. POLARITY IS, THEREFORE, NOT APPLICABLE TO THEM.

If there was only one form of War, to wit, the attack of the enemy, therefore no defence; or, in other words, if the attack was distinguished from the defence merely by the positive motive, which the one has and the other has not, but the methods of each were precisely one and the same: then in this sort of fight every advantage gained on the one side would be a corresponding disadvantage on the other, and true polarity would exist.

But action in War is divided into two forms, attack and defence, which, as we shall hereafter explain more particularly, are very different and of unequal strength. Polarity therefore lies in that to which both bear a relation, in the decision, but not in the attack or defence itself.

If the one Commander wishes the solution put off, the other must wish to hasten it, but only by the same form of action. If it is A's interest not to attack his enemy at present, but four weeks hence, then it is B's interest to be attacked, not four weeks hence, but at the present moment. This is the direct antagonism of interests, but it by no means follows that it would be for B's interest to attack A at once. That is plainly something totally different.

17. THE EFFECT OF POLARITY IS OFTEN DESTROYED BY THE SUPERIORITY OF THE DEFENCE OVER THE ATTACK, AND THUS THE SUSPENSION OF ACTION IN WAR IS EXPLAINED.

If the form of defence is stronger than that of offence, as we shall
hereafter show, the question arises, Is the advantage of a deferred
decision as great on the one side as the advantage of the defensive
form on the other? If it is not, then it cannot by its counter-weight
over-balance the latter, and thus influence the progress of the action
of the war. We see, therefore, that the impulsive force existing in the
polarity of interests may be lost in the difference between the strength
of the offensive and the defensive, and thereby become ineffectual.

If, therefore, that side for which the present is favourable, is too
weak to be able to dispense with the advantage of the defensive, he must
put up with the unfavourable prospects which the future holds out; for
it may still be better to fight a defensive battle in the unpromising
future than to assume the offensive or make peace at present. Now, being
convinced that the superiority of the defensive(*) (rightly understood)
is very great, and much greater than may appear at first sight, we
conceive that the greater number of those periods of inaction which
occur in war are thus explained without involving any contradiction.
The weaker the motives to action are, the more will those motives be
absorbed and neutralised by this difference between attack and defence,
the more frequently, therefore, will action in warfare be stopped, as
indeed experience teaches.

   (*) It must be remembered that all this antedates by some
   years the introduction of long-range weapons.

18 A SECOND GROUND CONSISTS IN THE IMPERFECT KNOWLEDGE OF CIRCUMSTANCES.

But there is still another cause which may stop action in War, viz., an
incomplete view of the situation. Each Commander can only fully know his
own position; that of his opponent can only be known to him by reports,
which are uncertain; he may, therefore, form a wrong judgment with
respect to it upon data of this description, and, in consequence of that
error, he may suppose that the power of taking the initiative rests with
his adversary when it lies really with himself. This want of perfect
insight might certainly just as often occasion an untimely action as
untimely inaction, and hence it would in itself no more contribute
to delay than to accelerate action in war. Still, it must always be
regarded as one of the natural causes which may bring action in War to a
standstill without involving a contradiction. But if we reflect how much
more we are inclined and induced to estimate the power of our opponents
too high than too low, because it lies in human nature to do so, we
shall admit that our imperfect insight into facts in general must
contribute very much to delay action in War, and to modify the
application of the principles pending our conduct.

The possibility of a standstill brings into the action of War a new
modification, inasmuch as it dilutes that action with the element
of time, checks the influence or sense of danger in its course, and
increases the means of reinstating a lost balance of force. The
greater the tension of feelings from which the War springs, the greater
therefore the energy with which it is carried on, so much the shorter
will be the periods of inaction; on the other hand, the weaker the
principle of warlike activity, the longer will be these periods: for
powerful motives increase the force of the will, and this, as we know,
is always a factor in the product of force.

19. FREQUENT PERIODS OF INACTION IN WAR REMOVE IT FURTHER FROM THE
ABSOLUTE, AND MAKE IT STILL MORE A CALCULATION OF PROBABILITIES.

But the slower the action proceeds in War, the more frequent and
longer the periods of inaction, so much the more easily can an error
be repaired; therefore, so much the bolder a General will be in his
calculations, so much the more readily will he keep them below the line
of the absolute, and build everything upon probabilities and conjecture.

Thus, according as the course of the War is more or less slow, more or
less time will be allowed for that which the nature of a concrete
case particularly requires, calculation of probability based on given
circumstances.

20. THEREFORE, THE ELEMENT OF CHANCE ONLY IS WANTING TO MAKE OF WAR A
GAME, AND IN THAT ELEMENT IT IS LEAST OF ALL DEFICIENT.

We see from the foregoing how much the objective nature of War makes
it a calculation of probabilities; now there is only one single element
still wanting to make it a game, and that element it certainly is
not without: it is chance. There is no human affair which stands so
constantly and so generally in close connection with chance as War.
But together with chance, the accidental, and along with it good luck,
occupy a great place in War.

21. WAR IS A GAME BOTH OBJECTIVELY AND SUBJECTIVELY.

If we now take a look at the subjective nature of War, that is to say,
at those conditions under which it is carried on, it will appear to us
still more like a game. Primarily the element in which the operations
of War are carried on is danger; but which of all the moral qualities is
the first in danger? COURAGE. Now certainly courage is quite compatible
with prudent calculation, but still they are things of quite a different
kind, essentially different qualities of the mind; on the other
hand, daring reliance on good fortune, boldness, rashness, are only
expressions of courage, and all these propensities of the mind look for
the fortuitous (or accidental), because it is their element.

We see, therefore, how, from the commencement, the absolute, the
mathematical as it is called, nowhere finds any sure basis in the
calculations in the Art of War; and that from the outset there is a play
of possibilities, probabilities, good and bad luck, which spreads about
with all the coarse and fine threads of its web, and makes War of all
branches of human activity the most like a gambling game.

22. HOW THIS ACCORDS BEST WITH THE HUMAN MIND IN GENERAL.

Although our intellect always feels itself urged towards clearness and
certainty, still our mind often feels itself attracted by uncertainty.
Instead of threading its way with the understanding along the narrow
path of philosophical investigations and logical conclusions, in order,
almost unconscious of itself, to arrive in spaces where it feels itself
a stranger, and where it seems to part from all well-known objects, it
prefers to remain with the imagination in the realms of chance and luck.
Instead of living yonder on poor necessity, it revels here in the wealth
of possibilities; animated thereby, courage then takes wings to itself,
and daring and danger make the element into which it launches itself as
a fearless swimmer plunges into the stream.

Shall theory leave it here, and move on, self-satisfied with absolute
conclusions and rules? Then it is of no practical use. Theory must also
take into account the human element; it must accord a place to courage,
to boldness, even to rashness. The Art of War has to deal with living
and with moral forces, the consequence of which is that it can never
attain the absolute and positive. There is therefore everywhere a margin
for the accidental, and just as much in the greatest things as in the
smallest. As there is room for this accidental on the one hand, so on
the other there must be courage and self-reliance in proportion to the
room available. If these qualities are forthcoming in a high degree,
the margin left may likewise be great. Courage and self-reliance are,
therefore, principles quite essential to War; consequently, theory
must only set up such rules as allow ample scope for all degrees and
varieties of these necessary and noblest of military virtues. In daring
there may still be wisdom, and prudence as well, only they are estimated

by a different standard of value.

23. WAR IS ALWAYS A SERIOUS MEANS FOR A SERIOUS OBJECT. ITS MORE
PARTICULAR DEFINITION.

Such is War; such the Commander who conducts it; such the theory which
rules it. But War is no pastime; no mere passion for venturing and
winning; no work of a free enthusiasm: it is a serious means for a
serious object. All that appearance which it wears from the varying hues
of fortune, all that it assimilates into itself of the oscillations of
passion, of courage, of imagination, of enthusiasm, are only particular
properties of this means.

The War of a community--of whole Nations, and particularly of civilised
Nations--always starts from a political condition, and is called forth
by a political motive. It is, therefore, a political act. Now if it was
a perfect, unrestrained, and absolute expression of force, as we had to
deduct it from its mere conception, then the moment it is called forth
by policy it would step into the place of policy, and as something quite
independent of it would set it aside, and only follow its own laws, just
as a mine at the moment of explosion cannot be guided into any
other direction than that which has been given to it by preparatory
arrangements. This is how the thing has really been viewed hitherto,
whenever a want of harmony between policy and the conduct of a War has
led to theoretical distinctions of the kind. But it is not so, and the
idea is radically false. War in the real world, as we have already seen,
is not an extreme thing which expends itself at one single discharge; it
is the operation of powers which do not develop themselves completely
in the same manner and in the same measure, but which at one time expand
sufficiently to overcome the resistance opposed by inertia or friction,
while at another they are too weak to produce an effect; it is
therefore, in a certain measure, a pulsation of violent force more or
less vehement, consequently making its discharges and exhausting its
powers more or less quickly--in other words, conducting more or less
quickly to the aim, but always lasting long enough to admit of influence
being exerted on it in its course, so as to give it this or
that direction, in short, to be subject to the will of a guiding
intelligence., if we reflect that War has its root in a political
object, then naturally this original motive which called it into
existence should also continue the first and highest consideration in
its conduct. Still, the political object is no despotic lawgiver on
that account; it must accommodate itself to the nature of the means, and
though changes in these means may involve modification in the political
objective, the latter always retains a prior right to consideration.
Policy, therefore, is interwoven with the whole action of War, and must
exercise a continuous influence upon it, as far as the nature of the
forces liberated by it will permit.

24. WAR IS A MERE CONTINUATION OF POLICY BY OTHER MEANS.

We see, therefore, that War is not merely a political act, but also
a real political instrument, a continuation of political commerce,
a carrying out of the same by other means. All beyond this which is
strictly peculiar to War relates merely to the peculiar nature of the
means which it uses. That the tendencies and views of policy shall not
be incompatible with these means, the Art of War in general and the
Commander in each particular case may demand, and this claim is truly
not a trifling one. But however powerfully this may react on political
views in particular cases, still it must always be regarded as only a
modification of them; for the political view is the object, War is the
means, and the means must always include the object in our conception.

25. DIVERSITY IN THE NATURE OF WARS.

The greater and the more powerful the motives of a War, the more it

affects the whole existence of a people. The more violent the excitement
which precedes the War, by so much the nearer will the War approach
to its abstract form, so much the more will it be directed to the
destruction of the enemy, so much the nearer will the military and
political ends coincide, so much the more purely military and less
political the War appears to be; but the weaker the motives and the
tensions, so much the less will the natural direction of the military
element--that is, force--be coincident with the direction which the
political element indicates; so much the more must, therefore, the War
become diverted from its natural direction, the political object diverge
from the aim of an ideal War, and the War appear to become political.

But, that the reader may not form any false conceptions, we must
here observe that by this natural tendency of War we only mean the
philosophical, the strictly logical, and by no means the tendency of
forces actually engaged in conflict, by which would be supposed to be
included all the emotions and passions of the combatants. No doubt in
some cases these also might be excited to such a degree as to be with
difficulty restrained and confined to the political road; but in most
cases such a contradiction will not arise, because by the existence
of such strenuous exertions a great plan in harmony therewith would
be implied. If the plan is directed only upon a small object, then the
impulses of feeling amongst the masses will be also so weak that these
masses will require to be stimulated rather than repressed.

26. THEY MAY ALL BE REGARDED AS POLITICAL ACTS.

Returning now to the main subject, although it is true that in one
kind of War the political element seems almost to disappear, whilst in
another kind it occupies a very prominent place, we may still affirm
that the one is as political as the other; for if we regard the State
policy as the intelligence of the personified State, then amongst
all the constellations in the political sky whose movements it has
to compute, those must be included which arise when the nature of
its relations imposes the necessity of a great War. It is only if we
understand by policy not a true appreciation of affairs in general,
but the conventional conception of a cautious, subtle, also dishonest
craftiness, averse from violence, that the latter kind of War may belong
more to policy than the first.

27. INFLUENCE OF THIS VIEW ON THE RIGHT UNDERSTANDING OF MILITARY
HISTORY, AND ON THE FOUNDATIONS OF THEORY.

We see, therefore, in the first place, that under all circumstances
War is to be regarded not as an independent thing, but as a political
instrument; and it is only by taking this point of view that we can
avoid finding ourselves in opposition to all military history. This is
the only means of unlocking the great book and making it intelligible.
Secondly, this view shows us how Wars must differ in character according
to the nature of the motives and circumstances from which they proceed.

Now, the first, the grandest, and most decisive act of judgment which
the Statesman and General exercises is rightly to understand in this
respect the War in which he engages, not to take it for something, or to
wish to make of it something, which by the nature of its relations it
is impossible for it to be. This is, therefore, the first, the most
comprehensive, of all strategical questions. We shall enter into this
more fully in treating of the plan of a War.

For the present we content ourselves with having brought the subject
up to this point, and having thereby fixed the chief point of view from
which War and its theory are to be studied.

28. RESULT FOR THEORY.

War is, therefore, not only chameleon-like in character, because it changes its colour in some degree in each particular case, but it is also, as a whole, in relation to the predominant tendencies which are in it, a wonderful trinity, composed of the original violence of its elements, hatred and animosity, which may be looked upon as blind instinct; of the play of probabilities and chance, which make it a free activity of the soul; and of the subordinate nature of a political instrument, by which it belongs purely to the reason.

The first of these three phases concerns more the people the second, more the General and his Army; the third, more the Government. The passions which break forth in War must already have a latent existence in the peoples. The range which the display of courage and talents shall get in the realm of probabilities and of chance depends on the particular characteristics of the General and his Army, but the political objects belong to the Government alone.

These three tendencies, which appear like so many different law-givers, are deeply rooted in the nature of the subject, and at the same time variable in degree. A theory which would leave any one of them out of account, or set up any arbitrary relation between them, would immediately become involved in such a contradiction with the reality, that it might be regarded as destroyed at once by that alone.

The problem is, therefore, that theory shall keep itself poised in a manner between these three tendencies, as between three points of attraction.

The way in which alone this difficult problem can be solved we shall examine in the book on the "Theory of War." In every case the conception of War, as here defined, will be the first ray of light which shows us the true foundation of theory, and which first separates the great masses and allows us to distinguish them from one another.

CHAPTER II. END AND MEANS IN WAR

HAVING in the foregoing chapter ascertained the complicated and variable nature of War, we shall now occupy ourselves in examining into the influence which this nature has upon the end and means in War.

If we ask, first of all, for the object upon which the whole effort of War is to be directed, in order that it may suffice for the attainment of the political object, we shall find that it is just as variable as are the political object and the particular circumstances of the War.

If, in the next place, we keep once more to the pure conception of War, then we must say that the political object properly lies out of its province, for if War is an act of violence to compel the enemy to fulfil our will, then in every case all depends on our overthrowing the enemy, that is, disarming him, and on that alone. This object, developed from abstract conceptions, but which is also the one aimed at in a great many cases in reality, we shall, in the first place, examine in this reality.

In connection with the plan of a campaign we shall hereafter examine more closely into the meaning of disarming a nation, but here we must at once draw a distinction between three things, which, as three general objects, comprise everything else within them. They are the MILITARY POWER, THE COUNTRY, and THE WILL OF THE ENEMY.

The military power must be destroyed, that is, reduced to such a state as not to be able to prosecute the War. This is the sense in which we wish to be understood hereafter, whenever we use the expression "destruction of the enemy's military power."

The country must be conquered, for out of the country a new military force may be formed.

But even when both these things are done, still the War, that is, the
hostile feeling and action of hostile agencies, cannot be considered as
at an end as long as the will of the enemy is not subdued also; that is,
its Government and its Allies must be forced into signing a peace, or
the people into submission; for whilst we are in full occupation of the
country, the War may break out afresh, either in the interior or through
assistance given by Allies. No doubt, this may also take place after a
peace, but that shows nothing more than that every War does not carry in
itself the elements for a complete decision and final settlement.

But even if this is the case, still with the conclusion of peace a
number of sparks are always extinguished which would have smouldered on
quietly, and the excitement of the passions abates, because all those
whose minds are disposed to peace, of which in all nations and under
all circumstances there is always a great number, turn themselves
away completely from the road to resistance. Whatever may take place
subsequently, we must always look upon the object as attained, and the
business of War as ended, by a peace.

As protection of the country is the primary object for which the
military force exists, therefore the natural order is, that first of all
this force should be destroyed, then the country subdued; and through
the effect of these two results, as well as the position we then hold,
the enemy should be forced to make peace. Generally the destruction of
the enemy's force is done by degrees, and in just the same measure the
conquest of the country follows immediately. The two likewise usually
react upon each other, because the loss of provinces occasions a
diminution of military force. But this order is by no means necessary,
and on that account it also does not always take place. The enemy's
Army, before it is sensibly weakened, may retreat to the opposite side
of the country, or even quite outside of it. In this case, therefore,
the greater part or the whole of the country is conquered.

But this object of War in the abstract, this final means of attaining
the political object in which all others are combined, the DISARMING THE
ENEMY, is rarely attained in practice and is not a condition necessary
to peace. Therefore it can in no wise be set up in theory as a law.
There are innumerable instances of treaties in which peace has been
settled before either party could be looked upon as disarmed; indeed,
even before the balance of power had undergone any sensible alteration.
Nay, further, if we look at the case in the concrete, then we must say
that in a whole class of cases, the idea of a complete defeat of the
enemy would be a mere imaginative flight, especially when the enemy is
considerably superior.

The reason why the object deduced from the conception of War is not
adapted in general to real War lies in the difference between the two,
which is discussed in the preceding chapter. If it was as pure theory
gives it, then a War between two States of very unequal military
strength would appear an absurdity; therefore impossible. At most, the
inequality between the physical forces might be such that it could be
balanced by the moral forces, and that would not go far with our present
social condition in Europe. Therefore, if we have seen Wars take place
between States of very unequal power, that has been the case because
there is a wide difference between War in reality and its original
conception.

There are two considerations which as motives may practically take
the place of inability to continue the contest. The first is the
improbability, the second is the excessive price, of success.

According to what we have seen in the foregoing chapter, War must always
set itself free from the strict law of logical necessity, and seek aid
from the calculation of probabilities; and as this is so much the more
the case, the more the War has a bias that way, from the circumstances
out of which it has arisen--the smaller its motives are, and the
excitement it has raised--so it is also conceivable how out of this

calculation of probabilities even motives to peace may arise. War does not, therefore, always require to be fought out until one party is overthrown; and we may suppose that, when the motives and passions are slight, a weak probability will suffice to move that side to which it is unfavourable to give way. Now, were the other side convinced of this beforehand, it is natural that he would strive for this probability only, instead of first wasting time and effort in the attempt to achieve the total destruction of the enemy's Army.

Still more general in its influence on the resolution to peace is the consideration of the expenditure of force already made, and further required. As War is no act of blind passion, but is dominated by the political object, therefore the value of that object determines the measure of the sacrifices by which it is to be purchased. This will be the case, not only as regards extent, but also as regards duration. As soon, therefore, as the required outlay becomes so great that the political object is no longer equal in value, the object must be given up, and peace will be the result.

We see, therefore, that in Wars where one side cannot completely disarm the other, the motives to peace on both sides will rise or fall on each side according to the probability of future success and the required outlay. If these motives were equally strong on both sides, they would meet in the centre of their political difference. Where they are strong on one side, they might be weak on the other. If their amount is only sufficient, peace will follow, but naturally to the advantage of that side which has the weakest motive for its conclusion. We purposely pass over here the difference which the POSITIVE and NEGATIVE character of the political end must necessarily produce practically; for although that is, as we shall hereafter show, of the highest importance, still we are obliged to keep here to a more general point of view, because the original political views in the course of the War change very much, and at last may become totally different, JUST BECAUSE THEY ARE DETERMINED BY RESULTS AND PROBABLE EVENTS.

Now comes the question how to influence the probability of success. In the first place, naturally by the same means which we use when the object is the subjugation of the enemy, by the destruction of his military force and the conquest of his provinces; but these two means are not exactly of the same import here as they would be in reference to that object. If we attack the enemy's Army, it is a very different thing whether we intend to follow up the first blow with a succession of others, until the whole force is destroyed, or whether we mean to content ourselves with a victory to shake the enemy's feeling of security, to convince him of our superiority, and to instil into him a feeling of apprehension about the future. If this is our object, we only go so far in the destruction of his forces as is sufficient. In like manner, the conquest, of the enemy's provinces is quite a different measure if the object is not the destruction of the enemy's Army. In the latter case the destruction of the Army is the real effectual action, and the taking of the provinces only a consequence of it; to take them before the Army had been defeated would always be looked upon only as a necessary evil. On the other hand, if our views are not directed upon the complete destruction of the enemy's force, and if we are sure that the enemy does not seek but fears to bring matters to a bloody decision, the taking possession of a weak or defenceless province is an advantage in itself, and if this advantage is of sufficient importance to make the enemy apprehensive about the general result, then it may also be regarded as a shorter road to peace.

But now we come upon a peculiar means of influencing the probability of the result without destroying the enemy's Army, namely, upon the expeditions which have a direct connection with political views. If there are any enterprises which are particularly likely to break up the enemy's alliances or make them inoperative, to gain new alliances for ourselves, to raise political powers in our own favour, &c. &c., then it is easy to conceive how much these may increase the probability of success, and become a shorter way towards our object than the routing of the enemy's forces.

The second question is how to act upon the enemy's expenditure in
strength, that is, to raise the price of success.

The enemy's outlay in strength lies in the WEAR AND TEAR of his forces,
consequently in the DESTRUCTION of them on our part, and in the LOSS of
PROVINCES, consequently the CONQUEST of them by us.

Here, again, on account of the various significations of these means, so
likewise it will be found that neither of them will be identical in its
signification in all cases if the objects are different. The smallness
in general of this difference must not cause us perplexity, for in
reality the weakest motives, the finest shades of difference, often
decide in favour of this or that method of applying force. Our only
business here is to show that, certain conditions being supposed,
the possibility of attaining our purpose in different ways is no
contradiction, absurdity, nor even error.

Besides these two means, there are three other peculiar ways of directly
increasing the waste of the enemy's force. The first is INVASION, that
is THE OCCUPATION OF THE ENEMY'S TERRITORY, NOT WITH A VIEW TO KEEPING
IT, but in order to levy contributions upon it, or to devastate it.

The immediate object here is neither the conquest of the enemy's
territory nor the defeat of his armed force, but merely to DO HIM DAMAGE
IN A GENERAL WAY. The second way is to select for the object of our
enterprises those points at which we can do the enemy most harm. Nothing
is easier to conceive than two different directions in which our force
may be employed, the first of which is to be preferred if our object is
to defeat the enemy's Army, while the other is more advantageous if the
defeat of the enemy is out of the question. According to the usual mode
of speaking, we should say that the first is primarily military, the
other more political. But if we take our view from the highest point,
both are equally military, and neither the one nor the other can be
eligible unless it suits the circumstances of the case. The third,
by far the most important, from the great number of cases which it
embraces, is the WEARING OUT of the enemy. We choose this expression not
only to explain our meaning in few words, but because it represents the
thing exactly, and is not so figurative as may at first appear. The idea
of wearing out in a struggle amounts in practice to A GRADUAL EXHAUSTION
OF THE PHYSICAL POWERS AND OF THE WILL BY THE LONG CONTINUANCE OF
EXERTION.

Now, if we want to overcome the enemy by the duration of the contest, we
must content ourselves with as small objects as possible, for it is in
the nature of the thing that a great end requires a greater expenditure
of force than a small one; but the smallest object that we can propose
to ourselves is simple passive resistance, that is a combat without any
positive view. In this way, therefore, our means attain their greatest
relative value, and therefore the result is best secured. How far now
can this negative mode of proceeding be carried? Plainly not to absolute
passivity, for mere endurance would not be fighting; and the defensive
is an activity by which so much of the enemy's power must be destroyed
that he must give up his object. That alone is what we aim at in each
single act, and therein consists the negative nature of our object.

No doubt this negative object in its single act is not so effective
as the positive object in the same direction would be, supposing it
successful; but there is this difference in its favour, that it succeeds
more easily than the positive, and therefore it holds out greater
certainty of success; what is wanting in the efficacy of its single
act must be gained through time, that is, through the duration of the
contest, and therefore this negative intention, which constitutes the
principle of the pure defensive, is also the natural means of overcoming
the enemy by the duration of the combat, that is of wearing him out.

Here lies the origin of that difference of OFFENSIVE and DEFENSIVE, the influence of which prevails throughout the whole province of War. We cannot at present pursue this subject further than to observe that from this negative intention are to be deduced all the advantages and all the stronger forms of combat which are on the side of the Defensive, and in which that philosophical-dynamic law which exists between the greatness and the certainty of success is realised. We shall resume the consideration of all this hereafter.

If then the negative purpose, that is the concentration of all the means into a state of pure resistance, affords a superiority in the contest, and if this advantage is sufficient to BALANCE whatever superiority in numbers the adversary may have, then the mere DURATION of the contest will suffice gradually to bring the loss of force on the part of the adversary to a point at which the political object can no longer be an equivalent, a point at which, therefore, he must give up the contest. We see then that this class of means, the wearing out of the enemy, includes the great number of cases in which the weaker resists the stronger.

Frederick the Great, during the Seven Years' War, was never strong enough to overthrow the Austrian monarchy; and if he had tried to do so after the fashion of Charles the Twelfth, he would inevitably have had to succumb himself. But after his skilful application of the system of husbanding his resources had shown the powers allied against him, through a seven years' struggle, that the actual expenditure of strength far exceeded what they had at first anticipated, they made peace.

We see then that there are many ways to one's object in War; that the complete subjugation of the enemy is not essential in every case; that the destruction of the enemy's military force, the conquest of the enemy's provinces, the mere occupation of them, the mere invasion of them--enterprises which are aimed directly at political objects--lastly, a passive expectation of the enemy's blow, are all means which, each in itself, may be used to force the enemy's will according as the peculiar circumstances of the case lead us to expect more from the one or the other. We could still add to these a whole category of shorter methods of gaining the end, which might be called arguments ad hominem. What branch of human affairs is there in which these sparks of individual spirit have not made their appearance, surmounting all formal considerations? And least of all can they fail to appear in War, where the personal character of the combatants plays such an important part, both in the cabinet and in the field. We limit ourselves to pointing this out, as it would be pedantry to attempt to reduce such influences into classes. Including these, we may say that the number of possible ways of reaching the object rises to infinity.

To avoid under-estimating these different short roads to one's purpose, either estimating them only as rare exceptions, or holding the difference which they cause in the conduct of War as insignificant, we must bear in mind the diversity of political objects which may cause a War--measure at a glance the distance which there is between a death struggle for political existence and a War which a forced or tottering alliance makes a matter of disagreeable duty. Between the two innumerable gradations occur in practice. If we reject one of these gradations in theory, we might with equal right reject the whole, which would be tantamount to shutting the real world completely out of sight.

These are the circumstances in general connected with the aim which we have to pursue in War; let us now turn to the means.

There is only one single means, it is the FIGHT. However diversified this may be in form, however widely it may differ from a rough vent of hatred and animosity in a hand-to-hand encounter, whatever number of things may introduce themselves which are not actual fighting, still it is always implied in the conception of War that all the effects manifested have their roots in the combat.

That this must always be so in the greatest diversity and complication

of the reality is proved in a very simple manner. All that takes place
in War takes place through armed forces, but where the forces of
War, i.e., armed men, are applied, there the idea of fighting must of
necessity be at the foundation.

All, therefore, that relates to forces of War--all that is connected
with their creation, maintenance, and application--belongs to military
activity.

Creation and maintenance are obviously only the means, whilst
application is the object.

The contest in War is not a contest of individual against individual,
but an organised whole, consisting of manifold parts; in this great
whole we may distinguish units of two kinds, the one determined by the
subject, the other by the object. In an Army the mass of combatants
ranges itself always into an order of new units, which again form
members of a higher order. The combat of each of these members forms,
therefore, also a more or less distinct unit. Further, the motive of the
fight; therefore its object forms its unit.

Now, to each of these units which we distinguish in the contest we
attach the name of combat.

If the idea of combat lies at the foundation of every application of
armed power, then also the application of armed force in general is
nothing more than the determining and arranging a certain number of
combats.

Every activity in War, therefore, necessarily relates to the combat
either directly or indirectly. The soldier is levied, clothed, armed,
exercised, he sleeps, eats, drinks, and marches, all MERELY TO FIGHT AT
THE RIGHT TIME AND PLACE.

If, therefore, all the threads of military activity terminate in the
combat, we shall grasp them all when we settle the order of the combats.
Only from this order and its execution proceed the effects, never
directly from the conditions preceding them. Now, in the combat all the
action is directed to the DESTRUCTION of the enemy, or rather of
HIS FIGHTING POWERS, for this lies in the conception of combat. The
destruction of the enemy's fighting power is, therefore, always the
means to attain the object of the combat.

This object may likewise be the mere destruction of the enemy's armed
force; but that is not by any means necessary, and it may be something
quite different. Whenever, for instance, as we have shown, the defeat of
the enemy is not the only means to attain the political object, whenever
there are other objects which may be pursued as the aim in a War, then
it follows of itself that such other objects may become the object of
particular acts of Warfare, and therefore also the object of combats.

But even those combats which, as subordinate acts, are in the strict
sense devoted to the destruction of the enemy's fighting force need not
have that destruction itself as their first object.

If we think of the manifold parts of a great armed force, of the number
of circumstances which come into activity when it is employed, then it
is clear that the combat of such a force must also require a manifold
organisation, a subordinating of parts and formation. There may and must
naturally arise for particular parts a number of objects which are not
themselves the destruction of the enemy's armed force, and which, while
they certainly contribute to increase that destruction, do so only in
an indirect manner. If a battalion is ordered to drive the enemy from
a rising ground, or a bridge, &c., then properly the occupation of any
such locality is the real object, the destruction of the enemy's armed
force which takes place only the means or secondary matter. If the enemy
can be driven away merely by a demonstration, the object is attained all
the same; but this hill or bridge is, in point of fact, only required as

a means of increasing the gross amount of loss inflicted on the enemy's armed force. It is the case on the field of battle, much more must it be so on the whole theatre of war, where not only one Army is opposed to another, but one State, one Nation, one whole country to another. Here the number of possible relations, and consequently possible combinations, is much greater, the diversity of measures increased, and by the gradation of objects, each subordinate to another the first means employed is further apart from the ultimate object.

It is therefore for many reasons possible that the object of a combat is not the destruction of the enemy's force, that is, of the force immediately opposed to us, but that this only appears as a means. But in all such cases it is no longer a question of complete destruction, for the combat is here nothing else but a measure of strength--has in itself no value except only that of the present result, that is, of its decision.

But a measuring of strength may be effected in cases where the opposing sides are very unequal by a mere comparative estimate. In such cases no fighting will take place, and the weaker will immediately give way.

If the object of a combat is not always the destruction of the enemy's forces therein engaged--and if its object can often be attained as well without the combat taking place at all, by merely making a resolve to fight, and by the circumstances to which this resolution gives rise--then that explains how a whole campaign may be carried on with great activity without the actual combat playing any notable part in it.

That this may be so military history proves by a hundred examples. How many of those cases can be justified, that is, without involving a contradiction and whether some of the celebrities who rose out of them would stand criticism, we shall leave undecided, for all we have to do with the matter is to show the possibility of such a course of events in War.

We have only one means in War--the battle; but this means, by the infinite variety of paths in which it may be applied, leads us into all the different ways which the multiplicity of objects allows of, so that we seem to have gained nothing; but that is not the case, for from this unity of means proceeds a thread which assists the study of the subject, as it runs through the whole web of military activity and holds it together.

But we have considered the destruction of the enemy's force as one of the objects which maybe pursued in War, and left undecided what relative importance should be given to it amongst other objects. In certain cases it will depend on circumstances, and as a general question we have left its value undetermined. We are once more brought back upon it, and we shall be able to get an insight into the value which must necessarily be accorded to it.

The combat is the single activity in War; in the combat the destruction of the enemy opposed to us is the means to the end; it is so even when the combat does not actually take place, because in that case there lies at the root of the decision the supposition at all events that this destruction is to be regarded as beyond doubt. It follows, therefore, that the destruction of the enemy's military force is the foundation-stone of all action in War, the great support of all combinations, which rest upon it like the arch on its abutments. All action, therefore, takes place on the supposition that if the solution by force of arms which lies at its foundation should be realised, it will be a favourable one. The decision by arms is, for all operations in War, great and small, what cash payment is in bill transactions. However remote from each other these relations, however seldom the realisation may take place, still it can never entirely fail to occur.

If the decision by arms lies at the foundation of all combinations, then it follows that the enemy can defeat each of them by gaining a victory on the field, not merely in the one on which our combination directly depends, but also in any other encounter, if it is only important enough; for every important decision by arms--that is, destruction of the enemy's forces--reacts upon all preceding it, because, like a liquid element, they tend to bring themselves to a level.

Thus, the destruction of the enemy's armed force appears, therefore, always as the superior and more effectual means, to which all others must give way.

It is, however, only when there is a supposed equality in all other conditions that we can ascribe to the destruction of the enemy's armed force the greater efficacy. It would, therefore, be a great mistake to draw the conclusion that a blind dash must always gain the victory over skill and caution. An unskilful attack would lead to the destruction of our own and not of the enemy's force, and therefore is not what is here meant. The superior efficacy belongs not to the MEANS but to the END, and we are only comparing the effect of one realised purpose with the other.

If we speak of the destruction of the enemy's armed force, we must expressly point out that nothing obliges us to confine this idea to the mere physical force; on the contrary, the moral is necessarily implied as well, because both in fact are interwoven with each other, even in the most minute details, and therefore cannot be separated. But it is just in connection with the inevitable effect which has been referred to, of a great act of destruction (a great victory) upon all other decisions by arms, that this moral element is most fluid, if we may use that expression, and therefore distributes itself the most easily through all the parts.

Against the far superior worth which the destruction of the enemy's armed force has over all other means stands the expense and risk of this means, and it is only to avoid these that any other means are taken. That these must be costly stands to reason, for the waste of our own military forces must, ceteris paribus, always be greater the more our aim is directed upon the destruction of the enemy's power.

The danger lies in this, that the greater efficacy which we seek recoils on ourselves, and therefore has worse consequences in case we fail of success.

Other methods are, therefore, less costly when they succeed, less dangerous when they fail; but in this is necessarily lodged the condition that they are only opposed to similar ones, that is, that the enemy acts on the same principle; for if the enemy should choose the way of a great decision by arms, OUR MEANS MUST ON THAT ACCOUNT BE CHANGED AGAINST OUR WILL, IN ORDER TO CORRESPOND WITH HIS. Then all depends on the issue of the act of destruction; but of course it is evident that, ceteris paribus, in this act we must be at a disadvantage in all respects because our views and our means had been directed in part upon other objects, which is not the case with the enemy. Two different objects of which one is not part, the other exclude each other, and therefore a force which may be applicable for the one may not serve for the other. If, therefore, one of two belligerents is determined to seek the great decision by arms, then he has a high probability of success, as soon as he is certain his opponent will not take that way, but follows a different object; and every one who sets before himself any such other aim only does so in a reasonable manner, provided he acts on the supposition that his adversary has as little intention as he has of resorting to the great decision by arms.

But what we have here said of another direction of views and forces relates only to other POSITIVE OBJECTS, which we may propose to ourselves in War, besides the destruction of the enemy's force, not by any means to the pure defensive, which may be adopted with a view thereby to exhaust the enemy's forces. In the pure defensive the positive object is wanting, and therefore, while on the defensive, our forces cannot at the same time be directed on other objects; they can only be employed to defeat the intentions of the enemy.

We have now to consider the opposite of the destruction of the enemy's armed force, that is to say, the preservation of our own. These two efforts always go together, as they mutually act and react on each other; they are integral parts of one and the same view, and we have only to ascertain what effect is produced when one or the other has the predominance. The endeavour to destroy the enemy's force has a positive object, and leads to positive results, of which the final aim is the conquest of the enemy. The preservation of our own forces has a negative object, leads therefore to the defeat of the enemy's intentions, that is to pure resistance, of which the final aim can be nothing more than to prolong the duration of the contest, so that the enemy shall exhaust himself in it.

The effort with a positive object calls into existence the act of destruction; the effort with the negative object awaits it.

How far this state of expectation should and may be carried we shall enter into more particularly in the theory of attack and defence, at the origin of which we again find ourselves. Here we shall content ourselves with saying that the awaiting must be no absolute endurance, and that in the action bound up with it the destruction of the enemy's armed force engaged in this conflict may be the aim just as well as anything else. It would therefore be a great error in the fundamental idea to suppose that the consequence of the negative course is that we are precluded from choosing the destruction of the enemy's military force as our object, and must prefer a bloodless solution. The advantage which the negative effort gives may certainly lead to that, but only at the risk of its not being the most advisable method, as that question is dependent on totally different conditions, resting not with ourselves but with our opponents. This other bloodless way cannot, therefore, be looked upon at all as the natural means of satisfying our great anxiety to spare our forces; on the contrary, when circumstances are not favourable, it would be the means of completely ruining them. Very many Generals have fallen into this error, and been ruined by it. The only necessary effect resulting from the superiority of the negative effort is the delay of the decision, so that the party acting takes refuge in that way, as it were, in the expectation of the decisive moment. The consequence of that is generally THE POSTPONEMENT OF THE ACTION as much as possible in time, and also in space, in so far as space is in connection with it. If the moment has arrived in which this can no longer be done without ruinous disadvantage, then the advantage of the negative must be considered as exhausted, and then comes forward unchanged the effort for the destruction of the enemy's force, which was kept back by a counterpoise, but never discarded.

We have seen, therefore, in the foregoing reflections, that there are many ways to the aim, that is, to the attainment of the political object; but that the only means is the combat, and that consequently everything is subject to a supreme law: which is the DECISION BY ARMS; that where this is really demanded by one, it is a redress which cannot be refused by the other; that, therefore, a belligerent who takes any other way must make sure that his opponent will not take this means of redress, or his cause may be lost in that supreme court; hence therefore the destruction of the enemy's armed force, amongst all the objects which can be pursued in War, appears always as the one which overrules all others.

What may be achieved by combinations of another kind in War we shall only learn in the sequel, and naturally only by degrees. We content ourselves here with acknowledging in general their possibility, as something pointing to the difference between the reality and the conception, and to the influence of particular circumstances. But we could not avoid showing at once that the BLOODY SOLUTION OF THE CRISIS, the effort for the destruction of the enemy's force, is the firstborn son of War. If when political objects are unimportant, motives weak, the

excitement of forces small, a cautious commander tries in all kinds
of ways, without great crises and bloody solutions, to twist himself
skilfully into a peace through the characteristic weaknesses of his
enemy in the field and in the Cabinet, we have no right to find
fault with him, if the premises on which he acts are well founded and
justified by success; still we must require him to remember that he only
travels on forbidden tracks, where the God of War may surprise him; that
he ought always to keep his eye on the enemy, in order that he may not
have to defend himself with a dress rapier if the enemy takes up a sharp
sword.

The consequences of the nature of War, how ends and means act in it, how
in the modifications of reality it deviates sometimes more, sometimes
less, from its strict original conception, fluctuating backwards and
forwards, yet always remaining under that strict conception as under a
supreme law: all this we must retain before us, and bear constantly
in mind in the consideration of each of the succeeding subjects, if we
would rightly comprehend their true relations and proper importance, and
not become involved incessantly in the most glaring contradictions with
the reality, and at last with our own selves.

CHAPTER III. THE GENIUS FOR WAR

EVERY special calling in life, if it is to be followed with success,
requires peculiar qualifications of understanding and soul. Where
these are of a high order, and manifest themselves by extraordinary
achievements, the mind to which they belong is termed GENIUS.

We know very well that this word is used in many significations which
are very different both in extent and nature, and that with many of
these significations it is a very difficult task to define the essence
of Genius; but as we neither profess to be philosopher nor grammarian,
we must be allowed to keep to the meaning usual in ordinary language,
and to understand by "genius" a very high mental capacity for certain
employments.

We wish to stop for a moment over this faculty and dignity of the mind,
in order to vindicate its title, and to explain more fully the meaning
of the conception. But we shall not dwell on that (genius) which has
obtained its title through a very great talent, on genius properly so
called, that is a conception which has no defined limits. What we have
to do is to bring under consideration every common tendency of the
powers of the mind and soul towards the business of War, the whole of
which common tendencies we may look upon as the ESSENCE OF MILITARY
GENIUS. We say "common," for just therein consists military genius,
that it is not one single quality bearing upon War, as, for instance,
courage, while other qualities of mind and soul are wanting or have a
direction which is unserviceable for War, but that it is AN HARMONIOUS
ASSOCIATION OF POWERS, in which one or other may predominate, but none
must be in opposition.

If every combatant required to be more or less endowed with military
genius, then our armies would be very weak; for as it implies a peculiar
bent of the intelligent powers, therefore it can only rarely be found
where the mental powers of a people are called into requisition and
trained in many different ways. The fewer the employments followed by a
Nation, the more that of arms predominates, so much the more prevalent
will military genius also be found. But this merely applies to its
prevalence, by no means to its degree, for that depends on the general
state of intellectual culture in the country. If we look at a wild,
warlike race, then we find a warlike spirit in individuals much more
common than in a civilised people; for in the former almost every
warrior possesses it, whilst in the civilised whole, masses are only
carried away by it from necessity, never by inclination. But amongst
uncivilised people we never find a really great General, and very seldom
what we can properly call a military genius, because that requires

a development of the intelligent powers which cannot be found in an uncivilised state. That a civilised people may also have a warlike tendency and development is a matter of course; and the more this is general, the more frequently also will military spirit be found in individuals in their armies. Now as this coincides in such case with the higher degree of civilisation, therefore from such nations have issued forth the most brilliant military exploits, as the Romans and the French have exemplified. The greatest names in these and in all other nations that have been renowned in War belong strictly to epochs of higher culture.

From this we may infer how great a share the intelligent powers have in superior military genius. We shall now look more closely into this point.

War is the province of danger, and therefore courage above all things is the first quality of a warrior.

Courage is of two kinds: first, physical courage, or courage in presence of danger to the person; and next, moral courage, or courage before responsibility, whether it be before the judgment-seat of external authority, or of the inner power, the conscience. We only speak here of the first.

Courage before danger to the person, again, is of two kinds. First, it may be indifference to danger, whether proceeding from the organism of the individual, contempt of death, or habit: in any of these cases it is to be regarded as a permanent condition.

Secondly, courage may proceed from positive motives, such as personal pride, patriotism, enthusiasm of any kind. In this case courage is not so much a normal condition as an impulse.

We may conceive that the two kinds act differently. The first kind is more certain, because it has become a second nature, never forsakes the man; the second often leads him farther. In the first there is more of firmness, in the second, of boldness. The first leaves the judgment cooler, the second raises its power at times, but often bewilders it. The two combined make up the most perfect kind of courage.

War is the province of physical exertion and suffering. In order not to be completely overcome by them, a certain strength of body and mind is required, which, either natural or acquired, produces indifference to them. With these qualifications, under the guidance of simply a sound understanding, a man is at once a proper instrument for war; and these are the qualifications so generally to be met with amongst wild and half-civilised tribes. If we go further in the demands which War makes on it, then we find the powers of the understanding predominating. War is the province of uncertainty: three-fourths of those things upon which action in War must be calculated, are hidden more or less in the clouds of great uncertainty. Here, then, above all a fine and penetrating mind is called for, to search out the truth by the tact of its judgment.

An average intellect may, at one time, perhaps hit upon this truth by accident; an extraordinary courage, at another, may compensate for the want of this tact; but in the majority of cases the average result will always bring to light the deficient understanding.

War is the province of chance. In no sphere of human activity is such a margin to be left for this intruder, because none is so much in constant contact with him on all sides. He increases the uncertainty of every circumstance, and deranges the course of events.

From this uncertainty of all intelligence and suppositions, this continual interposition of chance, the actor in War constantly finds things different from his expectations; and this cannot fail to have an influence on his plans, or at least on the presumptions connected with these plans. If this influence is so great as to render the pre-determined plan completely nugatory, then, as a rule, a new one must be substituted in its place; but at the moment the necessary data are often wanting for this, because in the course of action circumstances press for immediate decision, and allow no time to look about for fresh data, often not enough for mature consideration.

But it more often happens that the correction of one premise, and the knowledge of chance events which have arisen, are not sufficient to overthrow our plans completely, but only suffice to produce hesitation. Our knowledge of circumstances has increased, but our uncertainty, instead of having diminished, has only increased. The reason of this is, that we do not gain all our experience at once, but by degrees; thus our determinations continue to be assailed incessantly by fresh experience; and the mind, if we may use the expression, must always be "under arms."

Now, if it is to get safely through this perpetual conflict with the unexpected, two qualities are indispensable: in the first place an intellect which, even in the midst of this intense obscurity, is not without some traces of inner light, which lead to the truth, and then the courage to follow this faint light. The first is figuratively expressed by the French phrase coup d'oeil. The other is resolution. As the battle is the feature in War to which attention was originally chiefly directed, and as time and space are important elements in it, more particularly when cavalry with their rapid decisions were the chief arm, the idea of rapid and correct decision related in the first instance to the estimation of these two elements, and to denote the idea an expression was adopted which actually only points to a correct judgment by eye. Many teachers of the Art of War then gave this limited signification as the definition of coup d'oeil. But it is undeniable that all able decisions formed in the moment of action soon came to be understood by the expression, as, for instance, the hitting upon the right point of attack, &c. It is, therefore, not only the physical, but more frequently the mental eye which is meant in coup d'oeil. Naturally, the expression, like the thing, is always more in its place in the field of tactics: still, it must not be wanting in strategy, inasmuch as in it rapid decisions are often necessary. If we strip this conception of that which the expression has given it of the over-figurative and restricted, then it amounts simply to the rapid discovery of a truth which to the ordinary mind is either not visible at all or only becomes so after long examination and reflection.

Resolution is an act of courage in single instances, and if it becomes a characteristic trait, it is a habit of the mind. But here we do not mean courage in face of bodily danger, but in face of responsibility, therefore, to a certain extent against moral danger. This has been often called courage d'esprit, on the ground that it springs from the understanding; nevertheless, it is no act of the understanding on that account; it is an act of feeling. Mere intelligence is still not courage, for we often see the cleverest people devoid of resolution. The mind must, therefore, first awaken the feeling of courage, and then be guided and supported by it, because in momentary emergencies the man is swayed more by his feelings than his thoughts.

We have assigned to resolution the office of removing the torments of doubt, and the dangers of delay, when there are no sufficient motives for guidance. Through the unscrupulous use of language which is prevalent, this term is often applied to the mere propensity to daring, to bravery, boldness, or temerity. But, when there are SUFFICIENT MOTIVES in the man, let them be objective or subjective, true or false, we have no right to speak of his resolution; for, when we do so, we put ourselves in his place, and we throw into the scale doubts which did not exist with him.

Here there is no question of anything but of strength and weakness. We are not pedantic enough to dispute with the use of language about this little misapplication, our observation is only intended to remove wrong objections.

This resolution now, which overcomes the state of doubting, can only be
called forth by the intellect, and, in fact, by a peculiar tendency of
the same. We maintain that the mere union of a superior understanding
and the necessary feelings are not sufficient to make up resolution.
There are persons who possess the keenest perception for the most
difficult problems, who are also not fearful of responsibility, and yet
in cases of difficulty cannot come to a resolution. Their courage and
their sagacity operate independently of each other, do not give each
other a hand, and on that account do not produce resolution as a result.
The forerunner of resolution is an act of the mind making evident
the necessity of venturing, and thus influencing the will. This quite
peculiar direction of the mind, which conquers every other fear in man
by the fear of wavering or doubting, is what makes up resolution
in strong minds; therefore, in our opinion, men who have little
intelligence can never be resolute. They may act without hesitation
under perplexing circumstances, but then they act without reflection.
Now, of course, when a man acts without reflection he cannot be at
variance with himself by doubts, and such a mode of action may now
and then lead to the right point; but we say now as before, it is the
average result which indicates the existence of military genius. Should
our assertion appear extraordinary to any one, because he knows many a
resolute hussar officer who is no deep thinker, we must remind him that
the question here is about a peculiar direction of the mind, and not
about great thinking powers.

We believe, therefore, that resolution is indebted to a special
direction of the mind for its existence, a direction which belongs to
a strong head rather than to a brilliant one. In corroboration of this
genealogy of resolution we may add that there have been many instances
of men who have shown the greatest resolution in an inferior rank, and
have lost it in a higher position. While, on the one hand, they are
obliged to resolve, on the other they see the dangers of a wrong
decision, and as they are surrounded with things new to them, their
understanding loses its original force, and they become only the more
timid the more they become aware of the danger of the irresolution into
which they have fallen, and the more they have formerly been in the
habit of acting on the spur of the moment.

From the coup d'oeil and resolution we are naturally to speak of its
kindred quality, PRESENCE OF MIND, which in a region of the unexpected
like War must act a great part, for it is indeed nothing but a great
conquest over the unexpected. As we admire presence of mind in a
pithy answer to anything said unexpectedly, so we admire it in a ready
expedient on sudden danger. Neither the answer nor the expedient need be
in themselves extraordinary, if they only hit the point; for that which
as the result of mature reflection would be nothing unusual, therefore
insignificant in its impression on us, may as an instantaneous act of
the mind produce a pleasing impression. The expression "presence of
mind" certainly denotes very fitly the readiness and rapidity of the
help rendered by the mind.

Whether this noble quality of a man is to be ascribed more to the
peculiarity of his mind or to the equanimity of his feelings, depends
on the nature of the case, although neither of the two can be entirely
wanting. A telling repartee bespeaks rather a ready wit, a ready
expedient on sudden danger implies more particularly a well-balanced
mind.

If we take a general view of the four elements composing the atmosphere
in which War moves, of DANGER, PHYSICAL EFFORT, UNCERTAINTY, and CHANCE,
it is easy to conceive that a great force of mind and understanding is
requisite to be able to make way with safety and success amongst
such opposing elements, a force which, according to the different
modifications arising out of circumstances, we find termed by military
writers and annalists as ENERGY, FIRMNESS, STAUNCHNESS, STRENGTH OF MIND
AND CHARACTER. All these manifestations of the heroic nature might be
regarded as one and the same power of volition, modified according to

circumstances; but nearly related as these things are to each other,
still they are not one and the same, and it is desirable for us to
distinguish here a little more closely at least the action of the powers
of the soul in relation to them.

In the first place, to make the conception clear, it is essential to
observe that the weight, burden, resistance, or whatever it may be
called, by which that force of the soul in the General is brought to
light, is only in a very small measure the enemy's activity, the enemy's
resistance, the enemy's action directly. The enemy's activity only
affects the General directly in the first place in relation to his
person, without disturbing his action as Commander. If the enemy,
instead of two hours, resists for four, the Commander instead of
two hours is four hours in danger; this is a quantity which plainly
diminishes the higher the rank of the Commander. What is it for one in
the post of Commander-in-Chief? It is nothing.

Secondly, although the opposition offered by the enemy has a direct
effect on the Commander through the loss of means arising from prolonged
resistance, and the responsibility connected with that loss, and
his force of will is first tested and called forth by these anxious
considerations, still we maintain that this is not the heaviest burden
by far which he has to bear, because he has only himself to settle with.
All the other effects of the enemy's resistance act directly upon the
combatants under his command, and through them react upon him.

As long as his men full of good courage fight with zeal and spirit, it
is seldom necessary for the Chief to show great energy of purpose in the
pursuit of his object. But as soon as difficulties arise--and that must
always happen when great results are at stake--then things no longer
move on of themselves like a well-oiled machine, the machine itself then
begins to offer resistance, and to overcome this the Commander must have
a great force of will. By this resistance we must not exactly suppose
disobedience and murmurs, although these are frequent enough with
particular individuals; it is the whole feeling of the dissolution of
all physical and moral power, it is the heartrending sight of the bloody
sacrifice which the Commander has to contend with in himself, and
then in all others who directly or indirectly transfer to him their
impressions, feelings, anxieties, and desires. As the forces in one
individual after another become prostrated, and can no longer be excited
and supported by an effort of his own will, the whole inertia of the
mass gradually rests its weight on the will of the Commander: by the
spark in his breast, by the light of his spirit, the spark of purpose,
the light of hope, must be kindled afresh in others: in so far only
as he is equal to this, he stands above the masses and continues to be
their master; whenever that influence ceases, and his own spirit is
no longer strong enough to revive the spirit of all others, the masses
drawing him down with them sink into the lower region of animal nature,
which shrinks from danger and knows not shame. These are the weights
which the courage and intelligent faculties of the military Commander
have to overcome if he is to make his name illustrious. They increase
with the masses, and therefore, if the forces in question are to
continue equal to the burden, they must rise in proportion to the height
of the station.

Energy in action expresses the strength of the motive through which the
action is excited, let the motive have its origin in a conviction of
the understanding, or in an impulse. But the latter can hardly ever be
wanting where great force is to show itself.

Of all the noble feelings which fill the human heart in the exciting
tumult of battle, none, we must admit, are so powerful and constant
as the soul's thirst for honour and renown, which the German language
treats so unfairly and tends to depreciate by the unworthy associations
in the words Ehrgeiz (greed of honour) and Ruhmsucht (hankering after
glory). No doubt it is just in War that the abuse of these proud
aspirations of the soul must bring upon the human race the most shocking
outrages, but by their origin they are certainly to be counted amongst

the noblest feelings which belong to human nature, and in War they are
the vivifying principle which gives the enormous body a spirit. Although
other feelings may be more general in their influence, and many of
them--such as love of country, fanaticism, revenge, enthusiasm of every
kind--may seem to stand higher, the thirst for honour and renown still
remains indispensable. Those other feelings may rouse the great masses
in general, and excite them more powerfully, but they do not give
the Leader a desire to will more than others, which is an essential
requisite in his position if he is to make himself distinguished in it.
They do not, like a thirst for honour, make the military act specially
the property of the Leader, which he strives to turn to the best
account; where he ploughs with toil, sows with care, that he may reap
plentifully. It is through these aspirations we have been speaking of
in Commanders, from the highest to the lowest, this sort of energy,
this spirit of emulation, these incentives, that the action of armies is
chiefly animated and made successful. And now as to that which specially
concerns the head of all, we ask, Has there ever been a great
Commander destitute of the love of honour, or is such a character even
conceivable?

FIRMNESS denotes the resistance of the will in relation to the force of
a single blow, STAUNCHNESS in relation to a continuance of blows. Close
as is the analogy between the two, and often as the one is used in place
of the other, still there is a notable difference between them which
cannot be mistaken, inasmuch as firmness against a single powerful
impression may have its root in the mere strength of a feeling, but
staunchness must be supported rather by the understanding, for the
greater the duration of an action the more systematic deliberation is
connected with it, and from this staunchness partly derives its power.

If we now turn to STRENGTH OF MIND OR SOUL, then the first question is,
what are we to understand thereby?

Plainly it is not vehement expressions of feeling, nor easily excited
passions, for that would be contrary to all the usage of language,
but the power of listening to reason in the midst of the most intense
excitement, in the storm of the most violent passions. Should this power
depend on strength of understanding alone? We doubt it. The fact that
there are men of the greatest intellect who cannot command themselves
certainly proves nothing to the contrary, for we might say that it
perhaps requires an understanding of a powerful rather than of a
comprehensive nature; but we believe we shall be nearer the truth if
we assume that the power of submitting oneself to the control of the
understanding, even in moments of the most violent excitement of the
feelings, that power which we call SELF-COMMAND, has its root in the
heart itself. It is, in point of fact, another feeling, which in strong
minds balances the excited passions without destroying them; and it is
only through this equilibrium that the mastery of the understanding is
secured. This counterpoise is nothing but a sense of the dignity of man,
that noblest pride, that deeply-seated desire of the soul always to act
as a being endued with understanding and reason. We may therefore say
that a strong mind is one which does not lose its balance even under the
most violent excitement.

If we cast a glance at the variety to be observed in the human character
in respect to feeling, we find, first, some people who have very little
excitability, who are called phlegmatic or indolent.

Secondly, some very excitable, but whose feelings still never overstep
certain limits, and who are therefore known as men full of feeling, but
sober-minded.

Thirdly, those who are very easily roused, whose feelings blaze up
quickly and violently like gunpowder, but do not last.

Fourthly, and lastly, those who cannot be moved by slight causes, and
who generally are not to be roused suddenly, but only gradually; but
whose feelings become very powerful and are much more lasting. These are

men with strong passions, lying deep and latent.

This difference of character lies probably close on the confines of the physical powers which move the human organism, and belongs to that amphibious organisation which we call the nervous system, which appears to be partly material, partly spiritual. With our weak philosophy, we shall not proceed further in this mysterious field. But it is important for us to spend a moment over the effects which these different natures have on, action in War, and to see how far a great strength of mind is to be expected from them.

Indolent men cannot easily be thrown out of their equanimity, but we cannot certainly say there is strength of mind where there is a want of all manifestation of power.

At the same time, it is not to be denied that such men have a certain peculiar aptitude for War, on account of their constant equanimity. They often want the positive motive to action, impulse, and consequently activity, but they are not apt to throw things into disorder.

The peculiarity of the second class is that they are easily excited to act on trifling grounds, but in great matters they are easily overwhelmed. Men of this kind show great activity in helping an unfortunate individual, but by the distress of a whole Nation they are only inclined to despond, not roused to action.

Such people are not deficient in either activity or equanimity in War; but they will never accomplish anything great unless a great intellectual force furnishes the motive, and it is very seldom that a strong, independent mind is combined with such a character.

Excitable, inflammable feelings are in themselves little suited for practical life, and therefore they are not very fit for War. They have certainly the advantage of strong impulses, but that cannot long sustain them. At the same time, if the excitability in such men takes the direction of courage, or a sense of honour, they may often be very useful in inferior positions in War, because the action in War over which commanders in inferior positions have control is generally of shorter duration. Here one courageous resolution, one effervescence of the forces of the soul, will often suffice. A brave attack, a soul-stirring hurrah, is the work of a few moments, whilst a brave contest on the battle-field is the work of a day, and a campaign the work of a year.

Owing to the rapid movement of their feelings, it is doubly difficult for men of this description to preserve equilibrium of the mind; therefore they frequently lose head, and that is the worst phase in their nature as respects the conduct of War. But it would be contrary to experience to maintain that very excitable spirits can never preserve a steady equilibrium--that is to say, that they cannot do so even under the strongest excitement. Why should they not have the sentiment of self-respect, for, as a rule, they are men of a noble nature? This feeling is seldom wanting in them, but it has not time to produce an effect. After an outburst they suffer most from a feeling of inward humiliation. If through education, self-observance, and experience of life, they have learned, sooner or later, the means of being on their guard, so that at the moment of powerful excitement they are conscious betimes of the counteracting force within their own breasts, then even such men may have great strength of mind.

Lastly, those who are difficult to move, but on that account susceptible of very deep feelings, men who stand in the same relation to the preceding as red heat to a flame, are the best adapted by means of their Titanic strength to roll away the enormous masses by which we may figuratively represent the difficulties which beset command in War. The effect of their feelings is like the movement of a great body, slower, but more irresistible.

Although such men are not so likely to be suddenly surprised by their feelings and carried away so as to be afterwards ashamed of themselves, like the preceding, still it would be contrary to experience to believe that they can never lose their equanimity, or be overcome by blind passion; on the contrary, this must always happen whenever the noble pride of self-control is wanting, or as often as it has not sufficient weight. We see examples of this most frequently in men of noble minds belonging to savage nations, where the low degree of mental cultivation favours always the dominance of the passions. But even amongst the most civilised classes in civilised States, life is full of examples of this kind--of men carried away by the violence of their passions, like the poacher of old chained to the stag in the forest.

We therefore say once more a strong mind is not one that is merely susceptible of strong excitement, but one which can maintain its serenity under the most powerful excitement, so that, in spite of the storm in the breast, the perception and judgment can act with perfect freedom, like the needle of the compass in the storm-tossed ship.

By the term STRENGTH OF CHARACTER, or simply CHARACTER, is denoted tenacity of conviction, let it be the result of our own or of others' views, and whether they are principles, opinions, momentary inspirations, or any kind of emanations of the understanding; but this kind of firmness certainly cannot manifest itself if the views themselves are subject to frequent change. This frequent change need not be the consequence of external influences; it may proceed from the continuous activity of our own mind, in which case it indicates a characteristic unsteadiness of mind. Evidently we should not say of a man who changes his views every moment, however much the motives of change may originate with himself, that he has character. Only those men, therefore, can be said to have this quality whose conviction is very constant, either because it is deeply rooted and clear in itself, little liable to alteration, or because, as in the case of indolent men, there is a want of mental activity, and therefore a want of motives to change; or lastly, because an explicit act of the will, derived from an imperative maxim of the understanding, refuses any change of opinion up to a certain point.

Now in War, owing to the many and powerful impressions to which the mind is exposed, and in the uncertainty of all knowledge and of all science, more things occur to distract a man from the road he has entered upon, to make him doubt himself and others, than in any other human activity.

The harrowing sight of danger and suffering easily leads to the feelings gaining ascendency over the conviction of the understanding; and in the twilight which surrounds everything a deep clear view is so difficult that a change of opinion is more conceivable and more pardonable. It is, at all times, only conjecture or guesses at truth which we have to act upon. This is why differences of opinion are nowhere so great as in War, and the stream of impressions acting counter to one's own convictions never ceases to flow. Even the greatest impassibility of mind is hardly proof against them, because the impressions are powerful in their nature, and always act at the same time upon the feelings.

When the discernment is clear and deep, none but general principles and views of action from a high standpoint can be the result; and on these principles the opinion in each particular case immediately under consideration lies, as it were, at anchor. But to keep to these results of bygone reflection, in opposition to the stream of opinions and phenomena which the present brings with it, is just the difficulty. Between the particular case and the principle there is often a wide space which cannot always be traversed on a visible chain of conclusions, and where a certain faith in self is necessary and a certain amount of scepticism is serviceable. Here often nothing else will help us but an imperative maxim which, independent of reflection, at once controls it: that maxim is, in all doubtful cases to adhere to the first opinion, and not to give it up until a clear conviction forces us to do so. We must firmly believe in the superior authority of well-tried maxims, and under the dazzling influence of momentary events not forget that their value is of an inferior stamp. By this preference which in doubtful cases we give to first convictions, by adherence to the same our actions acquire that stability and consistency which make up what is called character.

It is easy to see how essential a well-balanced mind is to strength of character; therefore men of strong minds generally have a great deal of character.

Force of character leads us to a spurious variety of it--OBSTINACY.

It is often very difficult in concrete cases to say where the one ends and the other begins; on the other hand, it does not seem difficult to determine the difference in idea.

Obstinacy is no fault of the understanding; we use the term as denoting a resistance against our better judgment, and it would be inconsistent to charge that to the understanding, as the understanding is the power of judgment. Obstinacy is A FAULT OF THE FEELINGS or heart. This inflexibility of will, this impatience of contradiction, have their origin only in a particular kind of egotism, which sets above every other pleasure that of governing both self and others by its own mind alone. We should call it a kind of vanity, were it not decidedly something better. Vanity is satisfied with mere show, but obstinacy rests upon the enjoyment of the thing.

We say, therefore, force of character degenerates into obstinacy whenever the resistance to opposing judgments proceeds not from better convictions or a reliance upon a trustworthy maxim, but from a feeling of opposition. If this definition, as we have already admitted, is of little assistance practically, still it will prevent obstinacy from being considered merely force of character intensified, whilst it is something essentially different--something which certainly lies close to it and is cognate to it, but is at the same time so little an intensification of it that there are very obstinate men who from want of understanding have very little force of character.

Having in these high attributes of a great military Commander made ourselves acquainted with those qualities in which heart and head co-operate, we now come to a speciality of military activity which perhaps may be looked upon as the most marked if it is not the most important, and which only makes a demand on the power of the mind without regard to the forces of feelings. It is the connection which exists between War and country or ground.

This connection is, in the first place, a permanent condition of War, for it is impossible to imagine our organised Armies effecting any operation otherwise than in some given space; it is, secondly, of the most decisive importance, because it modifies, at times completely alters, the action of all forces; thirdly, while on the one hand it often concerns the most minute features of locality, on the other it may apply to immense tracts of country.

In this manner a great peculiarity is given to the effect of this connection of War with country and ground. If we think of other occupations of man which have a relation to these objects, on horticulture, agriculture, on building houses and hydraulic works, on mining, on the chase, and forestry, they are all confined within very limited spaces which may be soon explored with sufficient exactness. But the Commander in War must commit the business he has in hand to a corresponding space which his eye cannot survey, which the keenest zeal cannot always explore, and with which, owing to the constant changes taking place, he can also seldom become properly acquainted. Certainly the enemy generally is in the same situation; still, in the first place, the difficulty, although common to both, is not the less a difficulty, and he who by talent and practice overcomes it will have a great advantage on his side; secondly, this equality of the difficulty on both

sides is merely an abstract supposition which is rarely realised in the
particular case, as one of the two opponents (the defensive) usually
knows much more of the locality than his adversary.

This very peculiar difficulty must be overcome by a natural mental gift
of a special kind which is known by the--too restricted--term of
Orisinn sense of locality. It is the power of quickly forming a correct
geometrical idea of any portion of country, and consequently of being
able to find one's place in it exactly at any time. This is plainly
an act of the imagination. The perception no doubt is formed partly by
means of the physical eye, partly by the mind, which fills up what is
wanting with ideas derived from knowledge and experience, and out of the
fragments visible to the physical eye forms a whole; but that this whole
should present itself vividly to the reason, should become a picture, a
mentally drawn map, that this picture should be fixed, that the details
should never again separate themselves--all that can only be effected
by the mental faculty which we call imagination. If some great poet
or painter should feel hurt that we require from his goddess such an
office; if he shrugs his shoulders at the notion that a sharp gamekeeper
must necessarily excel in imagination, we readily grant that we only
speak here of imagination in a limited sense, of its service in a really
menial capacity. But, however slight this service, still it must be
the work of that natural gift, for if that gift is wanting, it would
be difficult to imagine things plainly in all the completeness of the
visible. That a good memory is a great assistance we freely allow, but
whether memory is to be considered as an independent faculty of the mind
in this case, or whether it is just that power of imagination which here
fixes these things better on the memory, we leave undecided, as in many
respects it seems difficult upon the whole to conceive these two mental
powers apart from each other.

That practice and mental acuteness have much to do with it is not to
be denied. Puysegur, the celebrated Quartermaster-General of the famous
Luxemburg, used to say that he had very little confidence in himself
in this respect at first, because if he had to fetch the parole from a
distance he always lost his way.

It is natural that scope for the exercise of this talent should increase
along with rank. If the hussar and rifleman in command of a patrol must
know well all the highways and byways, and if for that a few marks, a
few limited powers of observation, are sufficient, the Chief of an Army
must make himself familiar with the general geographical features of a
province and of a country; must always have vividly before his eyes
the direction of the roads, rivers, and hills, without at the same time
being able to dispense with the narrower "sense of locality" Orisinn.
No doubt, information of various kinds as to objects in general, maps,
books, memoirs, and for details the assistance of his Staff, are a great
help to him; but it is nevertheless certain that if he has himself a
talent for forming an ideal picture of a country quickly and distinctly,
it lends to his action an easier and firmer step, saves him from a
certain mental helplessness, and makes him less dependent on others.

If this talent then is to be ascribed to imagination, it is also almost
the only service which military activity requires from that erratic
goddess, whose influence is more hurtful than useful in other respects.

We think we have now passed in review those manifestations of the powers
of mind and soul which military activity requires from human nature.
Everywhere intellect appears as an essential co-operative force; and
thus we can understand how the work of War, although so plain and simple
in its effects, can never be conducted with distinguished success by
people without distinguished powers of the understanding.

When we have reached this view, then we need no longer look upon such a
natural idea as the turning an enemy's position, which has been done a
thousand times, and a hundred other similar conceptions, as the result
of a great effort of genius.

Certainly one is accustomed to regard the plain honest soldier as the
very opposite of the man of reflection, full of inventions and ideas, or
of the brilliant spirit shining in the ornaments of refined education of
every kind. This antithesis is also by no means devoid of truth; but it
does not show that the efficiency of the soldier consists only in his
courage, and that there is no particular energy and capacity of the
brain required in addition to make a man merely what is called a true
soldier. We must again repeat that there is nothing more common than to
hear of men losing their energy on being raised to a higher position,
to which they do not feel themselves equal; but we must also remind our
readers that we are speaking of pre-eminent services, of such as give
renown in the branch of activity to which they belong. Each grade of
command in War therefore forms its own stratum of requisite capacity of
fame and honour.

An immense space lies between a General--that is, one at the head of a
whole War, or of a theatre of War--and his Second in Command, for the
simple reason that the latter is in more immediate subordination to a
superior authority and supervision, consequently is restricted to a more
limited sphere of independent thought. This is why common opinion sees
no room for the exercise of high talent except in high places, and looks
upon an ordinary capacity as sufficient for all beneath: this is why
people are rather inclined to look upon a subordinate General grown grey
in the service, and in whom constant discharge of routine duties has
produced a decided poverty of mind, as a man of failing intellect, and,
with all respect for his bravery, to laugh at his simplicity. It is
not our object to gain for these brave men a better lot--that would
contribute nothing to their efficiency, and little to their happiness;
we only wish to represent things as they are, and to expose the error
of believing that a mere bravo without intellect can make himself
distinguished in War.

As we consider distinguished talents requisite for those who are to
attain distinction, even in inferior positions, it naturally follows
that we think highly of those who fill with renown the place of Second
in Command of an Army; and their seeming simplicity of character
as compared with a polyhistor, with ready men of business, or with
councillors of state, must not lead us astray as to the superior nature
of their intellectual activity. It happens sometimes that men import
the fame gained in an inferior position into a higher one, without in
reality deserving it in the new position; and then if they are not much
employed, and therefore not much exposed to the risk of showing their
weak points, the judgment does not distinguish very exactly what degree
of fame is really due to them; and thus such men are often the occasion
of too low an estimate being formed of the characteristics required to
shine in certain situations.

For each station, from the lowest upwards, to render distinguished
services in War, there must be a particular genius. But the title of
genius, history and the judgment of posterity only confer, in
general, on those minds which have shone in the highest rank, that of
Commanders-in-Chief. The reason is that here, in point of fact, the
demand on the reasoning and intellectual powers generally is much
greater.

To conduct a whole War, or its great acts, which we call campaigns, to
a successful termination, there must be an intimate knowledge of State
policy in its higher relations. The conduct of the War and the policy
of the State here coincide, and the General becomes at the same time the
Statesman.

We do not give Charles XII. the name of a great genius, because he could
not make the power of his sword subservient to a higher judgment and
philosophy--could not attain by it to a glorious object. We do not give
that title to Henry IV. (of France), because he did not live long
enough to set at rest the relations of different States by his military
activity, and to occupy himself in that higher field where noble
feelings and a chivalrous disposition have less to do in mastering the

enemy than in overcoming internal dissension.

In order that the reader may appreciate all that must be comprehended and judged of correctly at a glance by a General, we refer to the first chapter. We say the General becomes a Statesman, but he must not cease to be the General. He takes into view all the relations of the State on the one hand; on the other, he must know exactly what he can do with the means at his disposal.

As the diversity, and undefined limits, of all the circumstances bring a great number of factors into consideration in War, as the most of these factors can only be estimated according to probability, therefore, if the Chief of an Army does not bring to bear upon them a mind with an intuitive perception of the truth, a confusion of ideas and views must take place, in the midst of which the judgment will become bewildered. In this sense, Buonaparte was right when he said that many of the questions which come before a General for decision would make problems for a mathematical calculation not unworthy of the powers of Newton or Euler.

What is here required from the higher powers of the mind is a sense of unity, and a judgment raised to such a compass as to give the mind an extraordinary faculty of vision which in its range allays and sets aside a thousand dim notions which an ordinary understanding could only bring to light with great effort, and over which it would exhaust itself. But this higher activity of the mind, this glance of genius, would still not become matter of history if the qualities of temperament and character of which we have treated did not give it their support.

Truth alone is but a weak motive of action with men, and hence there is always a great difference between knowing and action, between science and art. The man receives the strongest impulse to action through the feelings, and the most powerful succour, if we may use the expression, through those faculties of heart and mind which we have considered under the terms of resolution, firmness, perseverance, and force of character.

If, however, this elevated condition of heart and mind in the General did not manifest itself in the general effects resulting from it, and could only be accepted on trust and faith, then it would rarely become matter of history.

All that becomes known of the course of events in War is usually very simple, and has a great sameness in appearance; no one on the mere relation of such events perceives the difficulties connected with them which had to be overcome. It is only now and again, in the memoirs of Generals or of those in their confidence, or by reason of some special historical inquiry directed to a particular circumstance, that a portion of the many threads composing the whole web is brought to light. The reflections, mental doubts, and conflicts which precede the execution of great acts are purposely concealed because they affect political interests, or the recollection of them is accidentally lost because they have been looked upon as mere scaffolding which had to be removed on the completion of the building.

If, now, in conclusion, without venturing upon a closer definition of the higher powers of the soul, we should admit a distinction in the intelligent faculties themselves according to the common ideas established by language, and ask ourselves what kind of mind comes closest to military genius, then a look at the subject as well as at experience will tell us that searching rather than inventive minds, comprehensive minds rather than such as have a special bent, cool rather than fiery heads, are those to which in time of War we should prefer to trust the welfare of our women and children, the honour and the safety of our fatherland.

CHAPTER IV. OF DANGER IN WAR

USUALLY before we have learnt what danger really is, we form an idea of it which is rather attractive than repulsive. In the intoxication of enthusiasm, to fall upon the enemy at the charge--who cares then about bullets and men falling? To throw oneself, blinded by excitement for a moment, against cold death, uncertain whether we or another shall escape him, and all this close to the golden gate of victory, close to the rich fruit which ambition thirsts for--can this be difficult? It will not be difficult, and still less will it appear so. But such moments, which, however, are not the work of a single pulse-beat, as is supposed, but rather like doctors' draughts, must be taken diluted and spoilt by mixture with time--such moments, we say, are but few.

Let us accompany the novice to the battle-field. As we approach, the thunder of the cannon becoming plainer and plainer is soon followed by the howling of shot, which attracts the attention of the inexperienced. Balls begin to strike the ground close to us, before and behind. We hasten to the hill where stands the General and his numerous Staff. Here the close striking of the cannon balls and the bursting of shells is so frequent that the seriousness of life makes itself visible through the youthful picture of imagination. Suddenly some one known to us falls--a shell strikes amongst the crowd and causes some involuntary movements--we begin to feel that we are no longer perfectly at ease and collected; even the bravest is at least to some degree confused. Now, a step farther into the battle which is raging before us like a scene in a theatre, we get to the nearest General of Division; here ball follows ball, and the noise of our own guns increases the confusion. From the General of Division to the Brigadier. He, a man of acknowledged bravery, keeps carefully behind a rising ground, a house, or a tree--a sure sign of increasing danger. Grape rattles on the roofs of the houses and in the fields; cannon balls howl over us, and plough the air in all directions, and soon there is a frequent whistling of musket balls. A step farther towards the troops, to that sturdy infantry which for hours has maintained its firmness under this heavy fire; here the air is filled with the hissing of balls which announce their proximity by a short sharp noise as they pass within an inch of the ear, the head, or the breast.

To add to all this, compassion strikes the beating heart with pity at the sight of the maimed and fallen. The young soldier cannot reach any of these different strata of danger without feeling that the light of reason does not move here in the same medium, that it is not refracted in the same manner as in speculative contemplation. Indeed, he must be a very extraordinary man who, under these impressions for the first time, does not lose the power of making any instantaneous decisions. It is true that habit soon blunts such impressions; in half in hour we begin to be more or less indifferent to all that is going on around us: but an ordinary character never attains to complete coolness and the natural elasticity of mind; and so we perceive that here again ordinary qualities will not suffice--a thing which gains truth, the wider the sphere of activity which is to be filled. Enthusiastic, stoical, natural bravery, great ambition, or also long familiarity with danger--much of all this there must be if all the effects produced in this resistant medium are not to fall far short of that which in the student's chamber may appear only the ordinary standard.

Danger in War belongs to its friction; a correct idea of its influence is necessary for truth of perception, and therefore it is brought under notice here.

CHAPTER V. OF BODILY EXERTION IN WAR

IF no one were allowed to pass an opinion on the events of War, except at a moment when he is benumbed by frost, sinking from heat and thirst, or dying with hunger and fatigue, we should certainly have fewer judgments correct *objectively; but they would be so, SUBJECTIVELY, at least; that is, they would contain in themselves the exact relation between the person giving the judgment and the object. We can perceive this by observing how modestly subdued, even spiritless and desponding, is the opinion passed upon the results of untoward events by those who have been eye-witnesses, but especially if they have been parties concerned. This is, according to our view, a criterion of the influence which bodily fatigue exercises, and of the allowance to be made for it in matters of opinion.

Amongst the many things in War for which no tariff can be fixed, bodily effort may be specially reckoned. Provided there is no waste, it is a coefficient of all the forces, and no one can tell exactly to what extent it may be carried. But what is remarkable is, that just as only a strong arm enables the archer to stretch the bowstring to the utmost extent, so also in War it is only by means of a great directing spirit that we can expect the full power latent in the troops to be developed. For it is one thing if an Army, in consequence of great misfortunes, surrounded with danger, falls all to pieces like a wall that has been thrown down, and can only find safety in the utmost exertion of its bodily strength; it is another thing entirely when a victorious Army, drawn on by proud feelings only, is conducted at the will of its Chief. The same effort which in the one case might at most excite our pity must in the other call forth our admiration, because it is much more difficult to sustain.

By this comes to light for the inexperienced eye one of those things which put fetters in the dark, as it were, on the action of the mind, and wear out in secret the powers of the soul.

Although here the question is strictly only respecting the extreme effort required by a Commander from his Army, by a leader from his followers, therefore of the spirit to demand it and of the art of getting it, still the personal physical exertion of Generals and of the Chief Commander must not be overlooked. Having brought the analysis of War conscientiously up to this point, we could not but take account also of the weight of this small remaining residue.

We have spoken here of bodily effort, chiefly because, like danger, it belongs to the fundamental causes of friction, and because its indefinite quantity makes it like an elastic body, the friction of which is well known to be difficult to calculate.

To check the abuse of these considerations, of such a survey of things which aggravate the difficulties of War, nature has given our judgment a guide in our sensibilities, just as an individual cannot with advantage refer to his personal deficiencies if he is insulted and ill-treated, but may well do so if he has successfully repelled the affront, or has fully revenged it, so no Commander or Army will lessen the impression of a disgraceful defeat by depicting the danger, the distress, the exertions, things which would immensely enhance the glory of a victory. Thus our feeling, which after all is only a higher kind of judgment, forbids us to do what seems an act of justice to which our judgment would be inclined.

CHAPTER VI. INFORMATION IN WAR

By the word "information" we denote all the knowledge which we have of the enemy and his country; therefore, in fact, the foundation of all our ideas and actions. Let us just consider the nature of this foundation, its want of trustworthiness, its changefulness, and we shall soon feel what a dangerous edifice War is, how easily it may fall to pieces and bury us in its ruins. For although it is a maxim in all books that we should trust only certain information, that we must be always suspicious, that is only a miserable book comfort, belonging to that description of knowledge in which writers of systems and compendiums take refuge for want of anything better to say.

Great part of the information obtained in War is contradictory, a still greater part is false, and by far the greatest part is of a doubtful character. What is required of an officer is a certain power of discrimination, which only knowledge of men and things and good judgment can give. The law of probability must be his guide. This is not a trifling difficulty even in respect of the first plans, which can be formed in the chamber outside the real sphere of War, but it is enormously increased when in the thick of War itself one report follows hard upon the heels of another; it is then fortunate if these reports in contradicting each other show a certain balance of probability, and thus themselves call forth a scrutiny. It is much worse for the inexperienced when accident does not render him this service, but one report supports another, confirms it, magnifies it, finishes off the picture with fresh touches of colour, until necessity in urgent haste forces from us a resolution which will soon be discovered to be folly, all those reports having been lies, exaggerations, errors, &c. &c. In a few words, most reports are false, and the timidity of men acts as a multiplier of lies and untruths. As a general rule, every one is more inclined to lend credence to the bad than the good. Every one is inclined to magnify the bad in some measure, and although the alarms which are thus propagated like the waves of the sea subside into themselves, still, like them, without any apparent cause they rise again. Firm in reliance on his own better convictions, the Chief must stand like a rock against which the sea breaks its fury in vain. The role is not easy; he who is not by nature of a buoyant disposition, or trained by experience in War, and matured in judgment, may let it be his rule to do violence to his own natural conviction by inclining from the side of fear to that of hope; only by that means will he be able to preserve his balance. This difficulty of seeing things correctly, which is one of the greatest sources of friction in War, makes things appear quite different from what was expected. The impression of the senses is stronger than the force of the ideas resulting from methodical reflection, and this goes so far that no important undertaking was ever yet carried out without the Commander having to subdue new doubts in himself at the time of commencing the execution of his work. Ordinary men who follow the suggestions of others become, therefore, generally undecided on the spot; they think that they have found circumstances different from what they had expected, and this view gains strength by their again yielding to the suggestions of others. But even the man who has made his own plans, when he comes to see things with his own eyes will often think he has done wrong. Firm reliance on self must make him proof against the seeming pressure of the moment; his first conviction will in the end prove true, when the foreground scenery which fate has pushed on to the stage of War, with its accompaniments of terrific objects, is drawn aside and the horizon extended. This is one of the great chasms which separate CONCEPTION from EXECUTION.

CHAPTER VII. FRICTION IN WAR

As long as we have no personal knowledge of War, we cannot conceive where those difficulties lie of which so much is said, and what that genius and those extraordinary mental powers required in a General have really to do. All appears so simple, all the requisite branches of knowledge appear so plain, all the combinations so unimportant, that in comparison with them the easiest problem in higher mathematics impresses us with a certain scientific dignity. But if we have seen War, all becomes intelligible; and still, after all, it is extremely difficult to describe what it is which brings about this change, to specify this invisible and completely efficient factor.

Everything is very simple in War, but the simplest thing is difficult. These difficulties accumulate and produce a friction which no man can imagine exactly who has not seen War, Suppose now a traveller, who towards evening expects to accomplish the two stages at the end of his day's journey, four or five leagues, with post-horses, on the high road--it is nothing. He arrives now at the last station but one, finds no horses, or very bad ones; then a hilly country, bad roads; it is a dark night, and he is glad when, after a great deal of trouble, he reaches the next station, and finds there some miserable accommodation. So in War, through the influence of an infinity of petty circumstances, which cannot properly be described on paper, things disappoint us, and we fall short of the mark. A powerful iron will overcomes this friction; it crushes the obstacles, but certainly the machine along with them. We shall often meet with this result. Like an obelisk towards which the principal streets of a town converge, the strong will of a proud spirit stands prominent and commanding in the middle of the Art of War.

Friction is the only conception which in a general way corresponds to that which distinguishes real War from War on paper. The military machine, the Army and all belonging to it, is in fact simple, and appears on this account easy to manage. But let us reflect that no part of it is in one piece, that it is composed entirely of individuals, each of which keeps up its own friction in all directions. Theoretically all sounds very well: the commander of a battalion is responsible for the execution of the order given; and as the battalion by its discipline is glued together into one piece, and the chief must be a man of acknowledged zeal, the beam turns on an iron pin with little friction. But it is not so in reality, and all that is exaggerated and false in such a conception manifests itself at once in War. The battalion always remains composed of a number of men, of whom, if chance so wills, the most insignificant is able to occasion delay and even irregularity. The danger which War brings with it, the bodily exertions which it requires, augment this evil so much that they may be regarded as the greatest causes of it.

This enormous friction, which is not concentrated, as in mechanics, at a few points, is therefore everywhere brought into contact with chance, and thus incidents take place upon which it was impossible to calculate, their chief origin being chance. As an instance of one such chance: the weather. Here the fog prevents the enemy from being discovered in time, a battery from firing at the right moment, a report from reaching the General; there the rain prevents a battalion from arriving at the right time, because instead of for three it had to march perhaps eight hours; the cavalry from charging effectively because it is stuck fast in heavy ground.

These are only a few incidents of detail by way of elucidation, that the reader may be able to follow the author, for whole volumes might be written on these difficulties. To avoid this, and still to give a clear conception of the host of small difficulties to be contended with in War, we might go on heaping up illustrations, if we were not afraid of being tiresome. But those who have already comprehended us will permit us to add a few more.

Activity in War is movement in a resistant medium. Just as a man immersed in water is unable to perform with ease and regularity the most natural and simplest movement, that of walking, so in War, with ordinary powers, one cannot keep even the line of mediocrity. This is the reason that the correct theorist is like a swimming master, who teaches on dry land movements which are required in the water, which must appear grotesque and ludicrous to those who forget about the water. This is also why theorists, who have never plunged in themselves, or who cannot deduce any generalities from their experience, are unpractical and even absurd, because they only teach what every one knows--how to walk.

Further, every War is rich in particular facts, while at the same time each is an unexplored sea, full of rocks which the General may have a suspicion of, but which he has never seen with his eye, and round which, moreover, he must steer in the night. If a contrary wind also springs up, that is, if any great accidental event declares itself adverse to him, then the most consummate skill, presence of mind, and energy are required, whilst to those who only look on from a distance all seems to proceed with the utmost ease. The knowledge of this friction is a chief

part of that so often talked of, experience in War, which is required
in a good General. Certainly he is not the best General in whose mind it
assumes the greatest dimensions, who is the most over-awed by it (this
includes that class of over-anxious Generals, of whom there are so many
amongst the experienced); but a General must be aware of it that he may
overcome it, where that is possible, and that he may not expect a degree
of precision in results which is impossible on account of this very
friction. Besides, it can never be learnt theoretically; and if it
could, there would still be wanting that experience of judgment which
is called tact, and which is always more necessary in a field full of
innumerable small and diversified objects than in great and decisive
cases, when one's own judgment may be aided by consultation with others.
Just as the man of the world, through tact of judgment which has become
habit, speaks, acts, and moves only as suits the occasion, so the
officer experienced in War will always, in great and small matters, at
every pulsation of War as we may say, decide and determine suitably to
the occasion. Through this experience and practice the idea comes to his
mind of itself that so and so will not suit. And thus he will not easily
place himself in a position by which he is compromised, which, if
it often occurs in War, shakes all the foundations of confidence and
becomes extremely dangerous.

It is therefore this friction, or what is so termed here, which makes
that which appears easy in War difficult in reality. As we proceed, we
shall often meet with this subject again, and it will hereafter become
plain that besides experience and a strong will, there are still many
other rare qualities of the mind required to make a man a consummate
General.

CHAPTER VIII. CONCLUDING REMARKS, BOOK I

THOSE things which as elements meet together in the atmosphere of War
and make it a resistant medium for every activity we have designated
under the terms danger, bodily effort (exertion), information, and
friction. In their impedient effects they may therefore be comprehended
again in the collective notion of a general friction. Now is there,
then, no kind of oil which is capable of diminishing this friction? Only
one, and that one is not always available at the will of the Commander
or his Army. It is the habituation of an Army to War.

Habit gives strength to the body in great exertion, to the mind in great
danger, to the judgment against first impressions. By it a valuable
circumspection is generally gained throughout every rank, from the
hussar and rifleman up to the General of Division, which facilitates the
work of the Chief Commander.

As the human eye in a dark room dilates its pupil, draws in the little
light that there is, partially distinguishes objects by degrees, and
at last knows them quite well, so it is in War with the experienced
soldier, whilst the novice is only met by pitch dark night.

Habituation to War no General can give his Army at once, and the camps
of manoeuvre (peace exercises) furnish but a weak substitute for it,
weak in comparison with real experience in War, but not weak in relation
to other Armies in which the training is limited to mere mechanical
exercises of routine. So to regulate the exercises in peace time as
to include some of these causes of friction, that the judgment,
circumspection, even resolution of the separate leaders may be brought
into exercise, is of much greater consequence than those believe who do
not know the thing by experience. It is of immense importance that the
soldier, high or low, whatever rank he has, should not have to encounter
in War those things which, when seen for the first time, set him in
astonishment and perplexity; if he has only met with them one single
time before, even by that he is half acquainted with them. This relates
even to bodily fatigues. They should be practised less to accustom the
body to them than the mind. In War the young soldier is very apt to

regard unusual fatigues as the consequence of faults, mistakes, and embarrassment in the conduct of the whole, and to become distressed and despondent as a consequence. This would not happen if he had been prepared for this beforehand by exercises in peace.

Another less comprehensive but still very important means of gaining habituation to war in time of peace is to invite into the service officers of foreign armies who have had experience in war. Peace seldom reigns over all Europe, and never in all quarters of the world. A State which has been long at peace should, therefore, always seek to procure some officers who have done good service at the different scenes of warfare, or to send there some of its own, that they may get a lesson in war.

However small the number of officers of this description may appear in proportion to the mass, still their influence is very sensibly felt.(*) Their experience, the bent of their genius, the stamp of their character, influence their subordinates and comrades; and besides that, if they cannot be placed in positions of superior command, they may always be regarded as men acquainted with the country, who may be questioned on many special occasions.

> (*) The War of 1870 furnishes a marked illustration. Von Moltke and von Goeben, not to mention many others, had both seen service in this manner, the former in Turkey and Syria, the latter in Spain--EDITOR.

BOOK II. ON THE THEORY OF WAR

CHAPTER I. BRANCHES OF THE ART OF WAR

WAR in its literal meaning is fighting, for fighting alone is the efficient principle in the manifold activity which in a wide sense is called war. But fighting is a trial of strength of the moral and physical forces by means of the latter. That the moral cannot be omitted is evident of itself, for the condition of the mind has always the most decisive influence on the forces employed in war.

The necessity of fighting very soon led men to special inventions to turn the advantage in it in their own favour: in consequence of these the mode of fighting has undergone great alterations; but in whatever way it is conducted its conception remains unaltered, and fighting is that which constitutes war.

The inventions have been from the first weapons and equipments for the individual combatants. These have to be provided and the use of them learnt before the war begins. They are made suitable to the nature of the fighting, consequently are ruled by it; but plainly the activity engaged in these appliances is a different thing from the fight itself; it is only the preparation for the combat, not the conduct of the same. That arming and equipping are not essential to the conception of fighting is plain, because mere wrestling is also fighting.

Fighting has determined everything appertaining to arms and equipment, and these in turn modify the mode of fighting; there is, therefore, a reciprocity of action between the two.

Nevertheless, the fight itself remains still an entirely special activity, more particularly because it moves in an entirely special element, namely, in the element of danger.

If, then, there is anywhere a necessity for drawing a line between two different activities, it is here; and in order to see clearly the importance of this idea, we need only just to call to mind how often

On War.txt
eminent personal fitness in one field has turned out nothing but the
most useless pedantry in the other.

It is also in no way difficult to separate in idea the one activity from
the other, if we look at the combatant forces fully armed and equipped
as a given means, the profitable use of which requires nothing more than
a knowledge of their general results.

The Art of War is therefore, in its proper sense, the art of making use
of the given means in fighting, and we cannot give it a better name than
the "Conduct of War." On the other hand, in a wider sense all activities
which have their existence on account of War, therefore the whole
creation of troops, that is levying them, arming, equipping, and
exercising them, belong to the Art of War.

To make a sound theory it is most essential to separate these two
activities, for it is easy to see that if every act of War is to begin
with the preparation of military forces, and to presuppose forces so
organised as a primary condition for conducting War, that theory will
only be applicable in the few cases to which the force available happens
to be exactly suited. If, on the other hand, we wish to have a theory
which shall suit most cases, and will not be wholly useless in any case,
it must be founded on those means which are in most general use, and in
respect to these only on the actual results springing from them.

The conduct of War is, therefore, the formation and conduct of the
fighting. If this fighting was a single act, there would be no necessity
for any further subdivision, but the fight is composed of a greater
or less number of single acts, complete in themselves, which we call
combats, as we have shown in the first chapter of the first book, and
which form new units. From this arises the totally different activities,
that of the FORMATION and CONDUCT of these single combats in themselves,
and the COMBINATION of them with one another, with a view to the
ultimate object of the War. The first is called TACTICS, the other
STRATEGY.

This division into tactics and strategy is now in almost general use,
and every one knows tolerably well under which head to place any
single fact, without knowing very distinctly the grounds on which the
classification is founded. But when such divisions are blindly adhered
to in practice, they must have some deep root. We have searched for this
root, and we might say that it is just the usage of the majority which
has brought us to it. On the other hand, we look upon the arbitrary,
unnatural definitions of these conceptions sought to be established by
some writers as not in accordance with the general usage of the terms.

According to our classification, therefore, tactics IS THE THEORY OF THE
USE OF MILITARY FORCES IN COMBAT. Strategy IS THE THEORY OF THE USE OF
COMBATS FOR THE OBJECT OF THE WAR.

The way in which the conception of a single, or independent combat, is
more closely determined, the conditions to which this unit is attached,
we shall only be able to explain clearly when we consider the combat; we
must content ourselves for the present with saying that in relation
to space, therefore in combats taking place at the same time, the unit
reaches just as far as PERSONAL COMMAND reaches; but in regard to time,
and therefore in relation to combats which follow each other in close
succession, it reaches to the moment when the crisis which takes place
in every combat is entirely passed.

That doubtful cases may occur, cases, for instance, in which several
combats may perhaps be regarded also as a single one, will not overthrow
the ground of distinction we have adopted, for the same is the case with
all grounds of distinction of real things which are differentiated by a
gradually diminishing scale. There may, therefore, certainly be acts of
activity in War which, without any alteration in the point of view,
may just as well be counted strategic as tactical; for example, very
extended positions resembling a chain of posts, the preparations for the

passage of a river at several points, &c.

Our classification reaches and covers only the USE OF THE MILITARY
FORCE. But now there are in War a number of activities which are
subservient to it, and still are quite different from it; sometimes
closely allied, sometimes less near in their affinity. All these
activities relate to the MAINTENANCE OF THE MILITARY FORCE. In the same
way as its creation and training precede its use, so its maintenance is
always a necessary condition. But, strictly viewed, all activities thus
connected with it are always to be regarded only as preparations for
fighting; they are certainly nothing more than activities which are very
close to the action, so that they run through the hostile act alternate
in importance with the use of the forces. We have therefore a right to
exclude them as well as the other preparatory activities from the Art of
War in its restricted sense, from the conduct of War properly so called;
and we are obliged to do so if we would comply with the first principle
of all theory, the elimination of all heterogeneous elements. Who would
include in the real "conduct of War" the whole litany of subsistence and
administration, because it is admitted to stand in constant reciprocal
action with the use of the troops, but is something essentially
different from it?

We have said, in the third chapter of our first book, that as the fight
or combat is the only directly effective activity, therefore the threads
of all others, as they end in it, are included in it. By this we meant
to say that to all others an object was thereby appointed which, in
accordance with the laws peculiar to themselves, they must seek to
attain. Here we must go a little closer into this subject.

The subjects which constitute the activities outside of the combat are
of various kinds.

The one part belongs, in one respect, to the combat itself, is identical
with it, whilst it serves in another respect for the maintenance of the
military force. The other part belongs purely to the subsistence, and
has only, in consequence of the reciprocal action, a limited influence
on the combats by its results. The subjects which in one respect belong
to the fighting itself are MARCHES, CAMPS, and CANTONMENTS, for they
suppose so many different situations of troops, and where troops are
supposed there the idea of the combat must always be present.

The other subjects, which only belong to the maintenance, are
SUBSISTENCE, CARE OF THE SICK, the SUPPLY AND REPAIR OF ARMS AND
EQUIPMENT.

Marches are quite identical with the use of the troops. The act of
marching in the combat, generally called manoeuvring, certainly does
not necessarily include the use of weapons, but it is so completely
and necessarily combined with it that it forms an integral part of that
which we call a combat. But the march outside the combat is nothing but
the execution of a strategic measure. By the strategic plan is settled
WHEN, WHERE, and WITH WHAT FORCES a battle is to be delivered--and to
carry that into execution the march is the only means.

The march outside of the combat is therefore an instrument of strategy,
but not on that account exclusively a subject of strategy, for as the
armed force which executes it may be involved in a possible combat at
any moment, therefore its execution stands also under tactical as
well as strategic rules. If we prescribe to a column its route on a
particular side of a river or of a branch of a mountain, then that is
a strategic measure, for it contains the intention of fighting on that
particular side of the hill or river in preference to the other, in case
a combat should be necessary during the march.

But if a column, instead of following the road through a valley, marches
along the parallel ridge of heights, or for the convenience of
marching divides itself into several columns, then these are tactical
arrangements, for they relate to the manner in which we shall use the

troops in the anticipated combat.

The particular order of march is in constant relation with readiness for combat, is therefore tactical in its nature, for it is nothing more than the first or preliminary disposition for the battle which may possibly take place.

As the march is the instrument by which strategy apportions its active elements, the combats, but these last often only appear by their results and not in the details of their real course, it could not fail to happen that in theory the instrument has often been substituted for the efficient principle. Thus we hear of a decisive skilful march, allusion being thereby made to those combat-combinations to which these marches led. This substitution of ideas is too natural and conciseness of expression too desirable to call for alteration, but still it is only a condensed chain of ideas in regard to which we must never omit to bear in mind the full meaning, if we would avoid falling into error.

We fall into an error of this description if we attribute to strategical combinations a power independent of tactical results. We read of marches and manoeuvres combined, the object attained, and at the same time not a word about combat, from which the conclusion is drawn that there are means in War of conquering an enemy without fighting. The prolific nature of this error we cannot show until hereafter.

But although a march can be regarded absolutely as an integral part of the combat, still there are in it certain relations which do not belong to the combat, and therefore are neither tactical nor strategic. To these belong all arrangements which concern only the accommodation of the troops, the construction of bridges, roads, &c. These are only conditions; under many circumstances they are in very close connection, and may almost identify themselves with the troops, as in building a bridge in presence of the enemy; but in themselves they are always activities, the theory of which does not form part of the theory of the conduct of War.

Camps, by which we mean every disposition of troops in concentrated, therefore in battle order, in contradistinction to cantonments or quarters, are a state of rest, therefore of restoration; but they are at the same time also the strategic appointment of a battle on the spot, chosen; and by the manner in which they are taken up they contain the fundamental lines of the battle, a condition from which every defensive battle starts; they are therefore essential parts of both strategy and tactics.

Cantonments take the place of camps for the better refreshment of the troops. They are therefore, like camps, strategic subjects as regards position and extent; tactical subjects as regards internal organisation, with a view to readiness to fight.

The occupation of camps and cantonments no doubt usually combines with the recuperation of the troops another object also, for example, the covering a district of country, the holding a position; but it can very well be only the first. We remind our readers that strategy may follow a great diversity of objects, for everything which appears an advantage may be the object of a combat, and the preservation of the instrument with which War is made must necessarily very often become the object of its partial combinations.

If, therefore, in such a case strategy ministers only to the maintenance of the troops, we are not on that account out of the field of strategy, for we are still engaged with the use of the military force, because every disposition of that force upon any point whatever of the theatre of War is such a use.

But if the maintenance of the troops in camp or quarters calls forth activities which are no employment of the armed force, such as the construction of huts, pitching of tents, subsistence and sanitary services in camps or quarters, then such belong neither to strategy nor tactics.

Even entrenchments, the site and preparation of which are plainly part of the order of battle, therefore tactical subjects, do not belong to the theory of the conduct of War so far as respects the execution of their construction the knowledge and skill required for such work being, in point of fact, qualities inherent in the nature of an organised Army; the theory of the combat takes them for granted.

Amongst the subjects which belong to the mere keeping up of an armed force, because none of the parts are identified with the combat, the victualling of the troops themselves comes first, as it must be done almost daily and for each individual. Thus it is that it completely permeates military action in the parts constituting strategy--we say parts constituting strategy, because during a battle the subsistence of troops will rarely have any influence in modifying the plan, although the thing is conceivable enough. The care for the subsistence of the troops comes therefore into reciprocal action chiefly with strategy, and there is nothing more common than for the leading strategic features of a campaign and War to be traced out in connection with a view to this supply. But however frequent and however important these views of supply may be, the subsistence of the troops always remains a completely different activity from the use of the troops, and the former has only an influence on the latter by its results.

The other branches of administrative activity which we have mentioned stand much farther apart from the use of the troops. The care of sick and wounded, highly important as it is for the good of an Army, directly affects it only in a small portion of the individuals composing it, and therefore has only a weak and indirect influence upon the use of the rest. The completing and replacing articles of arms and equipment, except so far as by the organism of the forces it constitutes a continuous activity inherent in them--takes place only periodically, and therefore seldom affects strategic plans.

We must, however, here guard ourselves against a mistake. In certain cases these subjects may be really of decisive importance. The distance of hospitals and depots of munitions may very easily be imagined as the sole cause of very important strategic decisions. We do not wish either to contest that point or to throw it into the shade. But we are at present occupied not with the particular facts of a concrete case, but with abstract theory; and our assertion therefore is that such an influence is too rare to give the theory of sanitary measures and the supply of munitions and arms an importance in theory of the conduct of War such as to make it worth while to include in the theory of the conduct of War the consideration of the different ways and systems which the above theories may furnish, in the same way as is certainly necessary in regard to victualling troops.

If we have clearly understood the results of our reflections, then the activities belonging to War divide themselves into two principal classes, into such as are only "preparations for War" and into the "War itself." This division must therefore also be made in theory.

The knowledge and applications of skill in the preparations for War are engaged in the creation, discipline, and maintenance of all the military forces; what general names should be given to them we do not enter into, but we see that artillery, fortification, elementary tactics, as they are called, the whole organisation and administration of the various armed forces, and all such things are included. But the theory of War itself occupies itself with the use of these prepared means for the object of the war. It needs of the first only the results, that is, the knowledge of the principal properties of the means taken in hand for use. This we call "The Art of War" in a limited sense, or "Theory of the Conduct of War," or "Theory of the Employment of Armed Forces," all of them denoting for us the same thing.

The present theory will therefore treat the combat as the real contest,
marches, camps, and cantonments as circumstances which are more or less
identical with it. The subsistence of the troops will only come into
consideration like OTHER GIVEN CIRCUMSTANCES in respect of its results,
not as an activity belonging to the combat.

The Art of War thus viewed in its limited sense divides itself again
into tactics and strategy. The former occupies itself with the form of
the separate combat, the latter with its use. Both connect themselves
with the circumstances of marches, camps, cantonments only through the
combat, and these circumstances are tactical or strategic according as
they relate to the form or to the signification of the battle.

No doubt there will be many readers who will consider superfluous this
careful separation of two things lying so close together as tactics and
strategy, because it has no direct effect on the conduct itself of War.
We admit, certainly that it would be pedantry to look for direct effects
on the field of battle from a theoretical distinction.

But the first business of every theory is to clear up conceptions and
ideas which have been jumbled together, and, we may say, entangled and
confused; and only when a right understanding is established, as to
names and conceptions, can we hope to progress with clearness and
facility, and be certain that author and reader will always see things
from the same point of view. Tactics and strategy are two activities
mutually permeating each other in time and space, at the same time
essentially different activities, the inner laws and mutual relations of
which cannot be intelligible at all to the mind until a clear conception
of the nature of each activity is established.

He to whom all this is nothing, must either repudiate all theoretical
consideration, OR HIS UNDERSTANDING HAS NOT AS YET BEEN PAINED by the
confused and perplexing ideas resting on no fixed point of view,
leading to no satisfactory result, sometimes dull, sometimes fantastic,
sometimes floating in vague generalities, which we are often obliged to
hear and read on the conduct of War, owing to the spirit of scientific
investigation having hitherto been little directed to these subjects.

CHAPTER II. ON THE THEORY OF WAR

1. THE FIRST CONCEPTION OF THE "ART OF WAR" WAS MERELY THE PREPARATION
OF THE ARMED FORCES.

FORMERLY by the term "Art of War," or "Science of War," nothing was
understood but the totality of those branches of knowledge and those
appliances of skill occupied with material things. The pattern
and preparation and the mode of using arms, the construction of
fortifications and entrenchments, the organism of an army and the
mechanism of its movements, were the subject; these branches of knowledge
and skill above referred to, and the end and aim of them all was the
establishment of an armed force fit for use in War. All this concerned
merely things belonging to the material world and a one-sided activity
only, and it was in fact nothing but an activity advancing by gradations
from the lower occupations to a finer kind of mechanical art. The
relation of all this to War itself was very much the same as the
relation of the art of the sword cutler to the art of using the sword.
The employment in the moment of danger and in a state of constant
reciprocal action of the particular energies of mind and spirit in the
direction proposed to them was not yet even mooted.

2. TRUE WAR FIRST APPEARS IN THE ART OF SIEGES.

In the art of sieges we first perceive a certain degree of guidance of
the combat, something of the action of the intellectual faculties upon
the material forces placed under their control, but generally only so

far that it very soon embodied itself again in new material forms, such as approaches, trenches, counter-approaches, batteries, &c., and every step which this action of the higher faculties took was marked by some such result; it was only the thread that was required on which to string these material inventions in order. As the intellect can hardly manifest itself in this kind of War, except in such things, so therefore nearly all that was necessary was done in that way.

3. THEN TACTICS TRIED TO FIND ITS WAY IN THE SAME DIRECTION.

Afterwards tactics attempted to give to the mechanism of its joints the character of a general disposition, built upon the peculiar properties of the instrument, which character leads indeed to the battle-field, but instead of leading to the free activity of mind, leads to an Army made like an automaton by its rigid formations and orders of battle, which, movable only by the word of command, is intended to unwind its activities like a piece of clockwork.

4. THE REAL CONDUCT OF WAR ONLY MADE ITS APPEARANCE INCIDENTALLY AND INCOGNITO.

The conduct of War properly so called, that is, a use of the prepared means adapted to the most special requirements, was not considered as any suitable subject for theory, but one which should be left to natural talents alone. By degrees, as War passed from the hand-to-hand encounters of the middle ages into a more regular and systematic form, stray reflections on this point also forced themselves into men's minds, but they mostly appeared only incidentally in memoirs and narratives, and in a certain measure incognito.

5. REFLECTIONS ON MILITARY EVENTS BROUGHT ABOUT THE WANT OF A THEORY.

As contemplation on War continually increased, and its history every day assumed more of a critical character, the urgent want appeared of the support of fixed maxims and rules, in order that in the controversies naturally arising about military events the war of opinions might be brought to some one point. This whirl of opinions, which neither revolved on any central pivot nor according to any appreciable laws, could not but be very distasteful to people's minds.

6. ENDEAVOURS TO ESTABLISH A POSITIVE THEORY.

There arose, therefore, an endeavour to establish maxims, rules, and even systems for the conduct of War. By this the attainment of a positive object was proposed, without taking into view the endless difficulties which the conduct of War presents in that respect. The conduct of War, as we have shown, has no definite limits in any direction, while every system has the circumscribing nature of a synthesis, from which results an irreconcileable opposition between such a theory and practice.

7. LIMITATION TO MATERIAL OBJECTS.

Writers on theory felt the difficulty of the subject soon enough, and thought themselves entitled to get rid of it by directing their maxims and systems only upon material things and a one-sided activity. Their aim was to reach results, as in the science for the preparation for War, entirely certain and positive, and therefore only to take into consideration that which could be made matter of calculation.

8. SUPERIORITY OF NUMBERS.

The superiority in numbers being a material condition, it was chosen from amongst all the factors required to produce victory, because it could be brought under mathematical laws through combinations of time and space. It was thought possible to leave out of sight all other circumstances, by supposing them to be equal on each side, and therefore to neutralise one another. This would have been very well if it had been done to gain a preliminary knowledge of this one factor, according to its relations, but to make it a rule for ever to consider superiority of numbers as the sole law; to see the whole secret of the Art of War in the formula, IN A CERTAIN TIME, AT A CERTAIN POINT, TO BRING UP SUPERIOR MASSES--was a restriction overruled by the force of realities.

9. VICTUALLING OF TROOPS.

By one theoretical school an attempt was made to systematise another material element also, by making the subsistence of troops, according to a previously established organism of the Army, the supreme legislator in the higher conduct of War. In this way certainly they arrived at definite figures, but at figures which rested on a number of arbitrary calculations, and which therefore could not stand the test of practical application.

10. BASE.

An ingenious author tried to concentrate in a single conception, that of a BASE, a whole host of objects amongst which sundry relations even with immaterial forces found their way in as well. The list comprised the subsistence of the troops, the keeping them complete in numbers and equipment, the security of communications with the home country, lastly, the security of retreat in case it became necessary; and, first of all, he proposed to substitute this conception of a base for all these things; then for the base itself to substitute its own length (extent); and, last of all, to substitute the angle formed by the army with this base: all this was done to obtain a pure geometrical result utterly useless. This last is, in fact, unavoidable, if we reflect that none of these substitutions could be made without violating truth and leaving out some of the things contained in the original conception. The idea of a base is a real necessity for strategy, and to have conceived it is meritorious; but to make such a use of it as we have depicted is completely inadmissible, and could not but lead to partial conclusions which have forced these theorists into a direction opposed to common sense, namely, to a belief in the decisive effect of the enveloping form of attack.

11. INTERIOR LINES.

As a reaction against this false direction, another geometrical principle, that of the so-called interior lines, was then elevated to the throne. Although this principle rests on a sound foundation, on the truth that the combat is the only effectual means in War, still it is, just on account of its purely geometrical nature, nothing but another case of one-sided theory which can never gain ascendency in the real world.

12. ALL THESE ATTEMPTS ARE OPEN TO OBJECTION.

All these attempts at theory are only to be considered in their analytical part as progress in the province of truth, but in their synthetical part, in their precepts and rules, they are quite unserviceable.

They strive after determinate quantities, whilst in War all is undetermined, and the calculation has always to be made with varying quantities.

They direct the attention only upon material forces, while the whole military action is penetrated throughout by intelligent forces and their effects.

They only pay regard to activity on one side, whilst War is a constant state of reciprocal action, the effects of which are mutual.

13. AS A RULE THEY EXCLUDE GENIUS.

All that was not attainable by such miserable philosophy, the offspring of partial views, lay outside the precincts of science--and was the field of genius, which RAISES ITSELF ABOVE RULES.

Pity the warrior who is contented to crawl about in this beggardom of rules, which are too bad for genius, over which it can set itself superior, over which it can perchance make merry! What genius does must be the best of all rules, and theory cannot do better than to show how and why it is so.

Pity the theory which sets itself in opposition to the mind! It cannot repair this contradiction by any humility, and the humbler it is so much the sooner will ridicule and contempt drive it out of real life.

14. THE DIFFICULTY OF THEORY AS SOON AS MORAL QUANTITIES COME INTO CONSIDERATION.

Every theory becomes infinitely more difficult from the moment that it touches on the province of moral quantities. Architecture and painting know quite well what they are about as long as they have only to do with matter; there is no dispute about mechanical or optical construction. But as soon as the moral activities begin their work, as soon as moral impressions and feelings are produced, the whole set of rules dissolves into vague ideas.

The science of medicine is chiefly engaged with bodily phenomena only; its business is with the animal organism, which, liable to perpetual change, is never exactly the same for two moments. This makes its practice very difficult, and places the judgment of the physician above his science; but how much more difficult is the case if a moral effect is added, and how much higher must we place the physician of the mind?

15. THE MORAL QUANTITIES MUST NOT BE EXCLUDED IN WAR.

But now the activity in War is never directed solely against matter; it is always at the same time directed against the intelligent force which gives life to this matter, and to separate the two from each other is impossible.

But the intelligent forces are only visible to the inner eye, and this is different in each person, and often different in the same person at different times.

As danger is the general element in which everything moves in War, it is also chiefly by courage, the feeling of one's own power, that the judgment is differently influenced. It is to a certain extent the crystalline lens through which all appearances pass before reaching the understanding.

And yet we cannot doubt that these things acquire a certain objective value simply through experience.

Every one knows the moral effect of a surprise, of an attack in flank or rear. Every one thinks less of the enemy's courage as soon as he turns his back, and ventures much more in pursuit than when pursued. Every one judges of the enemy's General by his reputed talents, by his age and experience, and shapes his course accordingly. Every one casts a scrutinising glance at the spirit and feeling of his own and the enemy's troops. All these and similar effects in the province of the moral nature of man have established themselves by experience, are perpetually recurring, and therefore warrant our reckoning them as real quantities of their kind. What could we do with any theory which should leave them out of consideration?

Certainly experience is an indispensable title for these truths. With psychological and philosophical sophistries no theory, no General, should meddle.

16. PRINCIPAL DIFFICULTY OF A THEORY FOR THE CONDUCT OF WAR.

In order to comprehend clearly the difficulty of the proposition which is contained in a theory for the conduct of War, and thence to deduce the necessary characteristics of such a theory, we must take a closer view of the chief particulars which make up the nature of activity in War.

17. FIRST SPECIALITY.--MORAL FORCES AND THEIR EFFECTS. (HOSTILE FEELING.)

The first of these specialities consists in the moral forces and effects.

The combat is, in its origin, the expression of HOSTILE FEELING, but in our great combats, which we call Wars, the hostile feeling frequently resolves itself into merely a hostile VIEW, and there is usually no innate hostile feeling residing in individual against individual. Nevertheless, the combat never passes off without such feelings being brought into activity. National hatred, which is seldom wanting in our Wars, is a substitute for personal hostility in the breast of individual opposed to individual. But where this also is wanting, and at first no animosity of feeling subsists, a hostile feeling is kindled by the combat itself; for an act of violence which any one commits upon us by order of his superior, will excite in us a desire to retaliate and be revenged on him, sooner than on the superior power at whose command the act was done. This is human, or animal if we will; still it is so. We are very apt to regard the combat in theory as an abstract trial of strength, without any participation on the part of the feelings, and that is one of the thousand errors which theorists deliberately commit, because they do not see its consequences.

Besides that excitation of feelings naturally arising from the combat itself, there are others also which do not essentially belong to it, but which, on account of their relationship, easily unite with it--ambition, love of power, enthusiasm of every kind, &c. &c.

18. THE IMPRESSIONS OF DANGER. (COURAGE.)

Finally, the combat begets the element of danger, in which all the activities of War must live and move, like the bird in the air or the fish in the water. But the influences of danger all pass into the feelings, either directly--that is, instinctively--or through the medium of the understanding. The effect in the first case would be a desire to escape from the danger, and, if that cannot be done, fright and anxiety. If this effect does not take place, then it is COURAGE, which is a counterpoise to that instinct. Courage is, however, by no means an act of the understanding, but likewise a feeling, like fear; the latter looks to the physical preservation, courage to the moral preservation. Courage, then, is a nobler instinct. But because it is so, it will not allow itself to be used as a lifeless instrument, which produces its effects exactly according to prescribed measure. Courage is therefore no mere counterpoise to danger in order to neutralise the latter in its effects, but a peculiar power in itself.

19. EXTENT OF THE INFLUENCE OF DANGER.

But to estimate exactly the influence of danger upon the principal
actors in War, we must not limit its sphere to the physical danger of
the moment. It dominates over the actor, not only by threatening him,
but also by threatening all entrusted to him, not only at the moment in
which it is actually present, but also through the imagination at all
other moments, which have a connection with the present; lastly, not
only directly by itself, but also indirectly by the responsibility which
makes it bear with tenfold weight on the mind of the chief actor. who
could advise, or resolve upon a great battle, without feeling his mind
more or less wrought up, or perplexed by, the danger and responsibility
which such a great act of decision carries in itself? We may say that
action in War, in so far as it is real action, not a mere condition, is
never out of the sphere of danger.

20. OTHER POWERS OF FEELING.

If we look upon these affections which are excited by hostility and
danger as peculiarly belonging to War, we do not, therefore, exclude
from it all others accompanying man in his life's journey. They will
also find room here frequently enough. Certainly we may say that many
a petty action of the passions is silenced in this serious business of
life; but that holds good only in respect to those acting in a lower
sphere, who, hurried on from one state of danger and exertion to
another, lose sight of the rest of the things of life, BECOME UNUSED
TO DECEIT, because it is of no avail with death, and so attain to
that soldierly simplicity of character which has always been the best
representative of the military profession. In higher regions it is
otherwise, for the higher a man's rank, the more he must look around
him; then arise interests on every side, and a manifold activity of
the passions of good and bad. Envy and generosity, pride and humility,
fierceness and tenderness, all may appear as active powers in this great
drama.

21. PECULIARITY OF MIND.

The peculiar characteristics of mind in the chief actor have, as well as
those of the feelings, a high importance. From an imaginative, flighty,
inexperienced head, and from a calm, sagacious understanding, different
things are to be expected.

22. FROM THE DIVERSITY IN MENTAL INDIVIDUALITIES ARISES THE DIVERSITY OF
WAYS LEADING TO THE END.

It is this great diversity in mental individuality, the influence of
which is to be supposed as chiefly felt in the higher ranks, because it
increases as we progress upwards, which chiefly produces the diversity
of ways leading to the end noticed by us in the first book, and which
gives, to the play of probabilities and chance, such an unequal share in
determining the course of events.

23. SECOND PECULIARITY.--LIVING REACTION.

The second peculiarity in War is the living reaction, and the reciprocal
action resulting therefrom. We do not here speak of the difficulty of
estimating that reaction, for that is included in the difficulty before
mentioned, of treating the moral powers as quantities; but of this, that
reciprocal action, by its nature, opposes anything like a regular
plan. The effect which any measure produces upon the enemy is the most
distinct of all the data which action affords; but every theory must
keep to classes (or groups) of phenomena, and can never take up the
really individual case in itself: that must everywhere be left to

judgment and talent. It is therefore natural that in a business such as War, which in its plan--built upon general circumstances--is so often thwarted by unexpected and singular accidents, more must generally be left to talent; and less use can be made of a THEORETICAL GUIDE than in any other.

24. THIRD PECULIARITY.--UNCERTAINTY OF ALL DATA.

Lastly, the great uncertainty of all data in War is a peculiar difficulty, because all action must, to a certain extent, be planned in a mere twilight, which in addition not unfrequently--like the effect of a fog or moonshine--gives to things exaggerated dimensions and an unnatural appearance.

What this feeble light leaves indistinct to the sight talent must discover, or must be left to chance. It is therefore again talent, or the favour of fortune, on which reliance must be placed, for want of objective knowledge.

25. POSITIVE THEORY IS IMPOSSIBLE.

With materials of this kind we can only say to ourselves that it is a sheer impossibility to construct for the Art of War a theory which, like a scaffolding, shall ensure to the chief actor an external support on all sides. In all those cases in which he is thrown upon his talent he would find himself away from this scaffolding of theory and in opposition to it, and, however many-sided it might be framed, the same result would ensue of which we spoke when we said that talent and genius act beyond the law, and theory is in opposition to reality.

26. MEANS LEFT BY WHICH A THEORY IS POSSIBLE (THE DIFFICULTIES ARE NOT EVERYWHERE EQUALLY GREAT).

Two means present themselves of getting out of this difficulty. In the first place, what we have said of the nature of military action in general does not apply in the same manner to the action of every one, whatever may be his standing. In the lower ranks the spirit of self-sacrifice is called more into request, but the difficulties which the understanding and judgment meet with are infinitely less. The field of occurrences is more confined. Ends and means are fewer in number. Data more distinct; mostly also contained in the actually visible. But the higher we ascend the more the difficulties increase, until in the Commander-in-Chief they reach their climax, so that with him almost everything must be left to genius.

Further, according to a division of the subject in AGREEMENT WITH ITS NATURE, the difficulties are not everywhere the same, but diminish the more results manifest themselves in the material world, and increase the more they pass into the moral, and become motives which influence the will. Therefore it is easier to determine, by theoretical rules, the order and conduct of a battle, than the use to be made of the battle itself. Yonder physical weapons clash with each other, and although mind is not wanting therein, matter must have its rights. But in the effects to be produced by battles when the material results become motives, we have only to do with the moral nature. In a word, it is easier to make a theory for TACTICS than for STRATEGY.

27. THEORY MUST BE OF THE NATURE OF OBSERVATIONS NOT OF DOCTRINE.

The second opening for the possibility of a theory lies in the point of view that it does not necessarily require to be a DIRECTION for action. As a general rule, whenever an ACTIVITY is for the most part occupied with the same objects over and over again, with the same ends and means, although there may be trifling alterations and a corresponding number of

varieties of combination, such things are capable of becoming a subject
of study for the reasoning faculties. But such study is just the most
essential part of every THEORY, and has a peculiar title to that name.
It is an analytical investigation of the subject that leads to an exact
knowledge; and if brought to bear on the results of experience, which in
our case would be military history, to a thorough familiarity with it.
The nearer theory attains the latter object, so much the more it passes
over from the objective form of knowledge into the subjective one of
skill in action; and so much the more, therefore, it will prove itself
effective when circumstances allow of no other decision but that of
personal talents; it will show its effects in that talent itself. If
theory investigates the subjects which constitute War; if it separates
more distinctly that which at first sight seems amalgamated; if it
explains fully the properties of the means; if it shows their probable
effects; if it makes evident the nature of objects; if it brings to
bear all over the field of War the light of essentially critical
investigation--then it has fulfilled the chief duties of its province.
It becomes then a guide to him who wishes to make himself acquainted
with War from books; it lights up the whole road for him, facilitates
his progress, educates his judgment, and shields him from error.

If a man of expertness spends half his life in the endeavour to clear up
an obscure subject thoroughly, he will probably know more about it than
a person who seeks to master it in a short time. Theory is instituted
that each person in succession may not have to go through the same
labour of clearing the ground and toiling through his subject, but may
find the thing in order, and light admitted on it. It should educate
the mind of the future leader in War, or rather guide him in his
self-instruction, but not accompany him to the field of battle; just
as a sensible tutor forms and enlightens the opening mind of a youth
without, therefore, keeping him in leading strings all through his life.

If maxims and rules result of themselves from the considerations which
theory institutes, if the truth accretes itself into that form of
crystal, then theory will not oppose this natural law of the mind; it
will rather, if the arch ends in such a keystone, bring it prominently
out; but so does this, only in order to satisfy the philosophical law
of reason, in order to show distinctly the point to which the lines all
converge, not in order to form out of it an algebraical formula for use
upon the battle-field; for even these maxims and rules serve more to
determine in the reflecting mind the leading outline of its habitual
movements than as landmarks indicating to it the way in the act of
execution.

28. BY THIS POINT OF VIEW THEORY BECOMES POSSIBLE, AND CEASES TO BE IN
CONTRADICTION TO PRACTICE.

Taking this point of view, there is a possibility afforded of a
satisfactory, that is, of a useful, theory of the conduct of War, never
coming into opposition with the reality, and it will only depend on
rational treatment to bring it so far into harmony with action that
between theory and practice there shall no longer be that absurd
difference which an unreasonable theory, in defiance of common sense,
has often produced, but which, just as often, narrow-mindedness and
ignorance have used as a pretext for giving way to their natural
incapacity.

29. THEORY THEREFORE CONSIDERS THE NATURE OF ENDS AND MEANS--ENDS AND
MEANS IN TACTICS.

Theory has therefore to consider the nature of the means and ends.

In tactics the means are the disciplined armed forces which are to carry
on the contest. The object is victory. The precise definition of this
conception can be better explained hereafter in the consideration of
the combat. Here we content ourselves by denoting the retirement of the

enemy from the field of battle as the sign of victory. By means of this victory strategy gains the object for which it appointed the combat, and which constitutes its special signification. This signification has certainly some influence on the nature of the victory. A victory which is intended to weaken the enemy's armed forces is a different thing from one which is designed only to put us in possession of a position. The signification of a combat may therefore have a sensible influence on the preparation and conduct of it, consequently will be also a subject of consideration in tactics.

30. CIRCUMSTANCES WHICH ALWAYS ATTEND THE APPLICATION OF THE MEANS.

As there are certain circumstances which attend the combat throughout, and have more or less influence upon its result, therefore these must be taken into consideration in the application of the armed forces.

These circumstances are the locality of the combat (ground), the time of day, and the weather.

31. LOCALITY.

The locality, which we prefer leaving for solution, under the head of "Country and Ground," might, strictly speaking, be without any influence at all if the combat took place on a completely level and uncultivated plain.

In a country of steppes such a case may occur, but in the cultivated countries of Europe it is almost an imaginary idea. Therefore a combat between civilised nations, in which country and ground have no influence, is hardly conceivable.

32. TIME OF DAY.

The time of day influences the combat by the difference between day and night; but the influence naturally extends further than merely to the limits of these divisions, as every combat has a certain duration, and great battles last for several hours. In the preparations for a great battle, it makes an essential difference whether it begins in the morning or the evening. At the same time, certainly many battles may be fought in which the question of the time of day is quite immaterial, and in the generality of cases its influence is only trifling.

33. WEATHER.

Still more rarely has the weather any decisive influence, and it is mostly only by fogs that it plays a part.

34. END AND MEANS IN STRATEGY.

Strategy has in the first instance only the victory, that is, the tactical result, as a means to its object, and ultimately those things which lead directly to peace. The application of its means to this object is at the same time attended by circumstances which have an influence thereon more or less.

35. CIRCUMSTANCES WHICH ATTEND THE APPLICATION OF THE MEANS OF STRATEGY.

These circumstances are country and ground, the former including the territory and inhabitants of the whole theatre of war; next the time of the day, and the time of the year as well; lastly, the weather, particularly any unusual state of the same, severe frost, &c.

36. THESE FORM NEW MEANS.

By bringing these things into combination with the results of a
combat, strategy gives this result--and therefore the combat--a special
signification, places before it a particular object. But when
this object is not that which leads directly to peace, therefore a
subordinate one, it is only to be looked upon as a means; and therefore
in strategy we may look upon the results of combats or victories, in all
their different significations, as means. The conquest of a position
is such a result of a combat applied to ground. But not only are the
different combats with special objects to be considered as means, but
also every higher aim which we may have in view in the combination of
battles directed on a common object is to be regarded as a means. A
winter campaign is a combination of this kind applied to the season.

There remain, therefore, as objects, only those things which may be
supposed as leading DIRECTLY to peace, Theory investigates all these
ends and means according to the nature of their effects and their mutual
relations.

37. STRATEGY DEDUCES ONLY FROM EXPERIENCE THE ENDS AND MEANS TO BE
EXAMINED.

The first question is, How does strategy arrive at a complete list of
these things? If there is to be a philosophical inquiry leading to an
absolute result, it would become entangled in all those difficulties
which the logical necessity of the conduct of War and its theory
exclude. It therefore turns to experience, and directs its attention on
those combinations which military history can furnish. In this manner,
no doubt, nothing more than a limited theory can be obtained, which
only suits circumstances such as are presented in history. But this
incompleteness is unavoidable, because in any case theory must either
have deduced from, or have compared with, history what it advances with
respect to things. Besides, this incompleteness in every case is more
theoretical than real.

One great advantage of this method is that theory cannot lose itself in
abstruse disquisitions, subtleties, and chimeras, but must always remain
practical.

38. HOW FAR THE ANALYSIS OF THE MEANS SHOULD BE CARRIED.

Another question is, How far should theory go in its analysis of the
means? Evidently only so far as the elements in a separate form present
themselves for consideration in practice. The range and effect of
different weapons is very important to tactics; their construction,
although these effects result from it, is a matter of indifference;
for the conduct of War is not making powder and cannon out of a given
quantity of charcoal, sulphur, and saltpetre, of copper and tin: the
given quantities for the conduct of War are arms in a finished state and
their effects. Strategy makes use of maps without troubling itself about
triangulations; it does not inquire how the country is subdivided into
departments and provinces, and how the people are educated and governed,
in order to attain the best military results; but it takes things as it
finds them in the community of European States, and observes where very
different conditions have a notable influence on War.

39. GREAT SIMPLIFICATION OF THE KNOWLEDGE REQUIRED.

That in this manner the number of subjects for theory is much
simplified, and the knowledge requisite for the conduct of War much
reduced, is easy to perceive. The very great mass of knowledge and
appliances of skill which minister to the action of War in general, and
which are necessary before an army fully equipped can take the field,
unite in a few great results before they are able to reach, in actual

War, the final goal of their activity; just as the streams of a country unite themselves in rivers before they fall into the sea. Only those activities emptying themselves directly into the sea of War have to be studied by him who is to conduct its operations.

40. THIS EXPLAINS THE RAPID GROWTH OF GREAT GENERALS, AND WHY A GENERAL IS NOT A MAN OF LEARNING.

This result of our considerations is in fact so necessary, any other would have made us distrustful of their accuracy. Only thus is explained how so often men have made their appearance with great success in War, and indeed in the higher ranks even in supreme Command, whose pursuits had been previously of a totally different nature; indeed how, as a rule, the most distinguished Generals have never risen from the very learned or really erudite class of officers, but have been mostly men who, from the circumstances of their position, could not have attained to any great amount of knowledge. On that account those who have considered it necessary or even beneficial to commence the education of a future General by instruction in all details have always been ridiculed as absurd pedants. It would be easy to show the injurious tendency of such a course, because the human mind is trained by the knowledge imparted to it and the direction given to its ideas. Only what is great can make it great; the little can only make it little, if the mind itself does not reject it as something repugnant.

41. FORMER CONTRADICTIONS.

Because this simplicity of knowledge requisite in War was not attended to, but that knowledge was always jumbled up with the whole impedimenta of subordinate sciences and arts, therefore the palpable opposition to the events of real life which resulted could not be solved otherwise than by ascribing it all to genius, which requires no theory and for which no theory could be prescribed.

42. ON THIS ACCOUNT ALL USE OF KNOWLEDGE WAS DENIED, AND EVERYTHING ASCRIBED TO NATURAL TALENTS.

People with whom common sense had the upper hand felt sensible of the immense distance remaining to be filled up between a genius of the highest order and a learned pedant; and they became in a manner free-thinkers, rejected all belief in theory, and affirmed the conduct of War to be a natural function of man, which he performs more or less well according as he has brought with him into the world more or less talent in that direction. It cannot be denied that these were nearer to the truth than those who placed a value on false knowledge: at the same time it may easily be seen that such a view is itself but an exaggeration. No activity of the human understanding is possible without a certain stock of ideas; but these are, for the greater part at least, not innate but acquired, and constitute his knowledge. The only question therefore is, of what kind should these ideas be; and we think we have answered it if we say that they should be directed on those things which man has directly to deal with in War.

43. THE KNOWLEDGE MUST BE MADE SUITABLE TO THE POSITION.

Inside this field itself of military activity, the knowledge required must be different according to the station of the Commander. It will be directed on smaller and more circumscribed objects if he holds an inferior, upon greater and more comprehensive ones if he holds a higher situation. There are Field Marshals who would not have shone at the head of a cavalry regiment, and vice versa.

44. THE KNOWLEDGE IN WAR IS VERY SIMPLE, BUT NOT, AT THE SAME TIME, VERY

EASY.

But although the knowledge in War is simple, that is to say directed to
so few subjects, and taking up those only in their final results, the
art of execution is not, on that account, easy. Of the difficulties to
which activity in War is subject generally, we have already spoken in
the first book; we here omit those things which can only be overcome by
courage, and maintain also that the activity of mind, is only simple,
and easy in inferior stations, but increases in difficulty with increase
of rank, and in the highest position, in that of Commander-in-Chief,
is to be reckoned among the most difficult which there is for the human
mind.

45. OF THE NATURE OF THIS KNOWLEDGE.

The Commander of an Army neither requires to be a learned explorer
of history nor a publicist, but he must be well versed in the higher
affairs of State; he must know, and be able to judge correctly of
traditional tendencies, interests at stake, the immediate questions at
issue, and the characters of leading persons; he need not be a close
observer of men, a sharp dissector of human character, but he must
know the character, the feelings, the habits, the peculiar faults and
inclinations of those whom he is to command. He need not understand
anything about the make of a carriage, or the harness of a battery
horse, but he must know how to calculate exactly the march of a column,
under different circumstances, according to the time it requires. These
are matters the knowledge of which cannot be forced out by an apparatus
of scientific formula and machinery: they are only to be gained by the
exercise of an accurate judgment in the observation of things and of
men, aided by a special talent for the apprehension of both.

The necessary knowledge for a high position in military action is
therefore distinguished by this, that by observation, therefore by study
and reflection, it is only to be attained through a special talent
which as an intellectual instinct understands how to extract from the
phenomena of life only the essence or spirit, as bees do the honey from
the flowers; and that it is also to be gained by experience of life as
well as by study and reflection. Life will never bring forth a Newton or
an Euler by its rich teachings, but it may bring forth great calculators
in War, such as Conde' or Frederick.

It is therefore not necessary that, in order to vindicate the
intellectual dignity of military activity, we should resort to untruth
and silly pedantry. There never has been a great and distinguished
Commander of contracted mind, but very numerous are the instances of men
who, after serving with the greatest distinction in inferior positions,
remained below mediocrity in the highest, from insufficiency of
intellectual capacity. That even amongst those holding the post of
Commander-in-Chief there may be a difference according to the degree of
their plenitude of power is a matter of course.

46. SCIENCE MUST BECOME ART.

Now we have yet to consider one condition which is more necessary for
the knowledge of the conduct of War than for any other, which is, that
it must pass completely into the mind and almost completely cease to be
something objective. In almost all other arts and occupations of life
the active agent can make use of truths which he has only learnt once,
and in the spirit and sense of which he no longer lives, and which he
extracts from dusty books. Even truths which he has in hand and uses
daily may continue something external to himself, If the architect takes
up a pen to settle the strength of a pier by a complicated calculation,
the truth found as a result is no emanation from his own mind. He had
first to find the data with labour, and then to submit these to an
operation of the mind, the rule for which he did not discover, the
necessity of which he is perhaps at the moment only partly conscious of,

but which he applies, for the most part, as if by mechanical dexterity.
But it is never so in War. The moral reaction, the ever-changeful form
of things, makes it necessary for the chief actor to carry in himself
the whole mental apparatus of his knowledge, that anywhere and at every
pulse-beat he may be capable of giving the requisite decision from
himself. Knowledge must, by this complete assimilation with his own
mind and life, be converted into real power. This is the reason
why everything seems so easy with men distinguished in War, and why
everything is ascribed to natural talent. We say natural talent, in
order thereby to distinguish it from that which is formed and matured by
observation and study.

We think that by these reflections we have explained the problem of a
theory of the conduct of War; and pointed out the way to its solution.

Of the two fields into which we have divided the conduct of War, tactics
and strategy, the theory of the latter contains unquestionably, as
before observed, the greatest difficulties, because the first is almost
limited to a circumscribed field of objects, but the latter, in the
direction of objects leading directly to peace, opens to itself
an unlimited field of possibilities. Since for the most part the
Commander-in-Chief has only to keep these objects steadily in view,
therefore the part of strategy in which he moves is also that which is
particularly subject to this difficulty.

Theory, therefore, especially where it comprehends the highest services,
will stop much sooner in strategy than in tactics at the simple
consideration of things, and content itself to assist the Commander to
that insight into things which, blended with his whole thought, makes
his course easier and surer, never forces him into opposition with
himself in order to obey an objective truth.

CHAPTER III. ART OR SCIENCE OF WAR

1.--USAGE STILL UNSETTLED

(POWER AND KNOWLEDGE. SCIENCE WHEN MERE KNOWING; ART, WHEN DOING, IS THE
OBJECT.)

THE choice between these terms seems to be still unsettled, and no one
seems to know rightly on what grounds it should be decided, and yet
the thing is simple. We have already said elsewhere that "knowing" is
something different from "doing." The two are so different that they
should not easily be mistaken the one for the other. The "doing" cannot
properly stand in any book, and therefore also Art should never be
the title of a book. But because we have once accustomed ourselves to
combine in conception, under the name of theory of Art, or simply
Art, the branches of knowledge (which may be separately pure sciences)
necessary for the practice of an Art, therefore it is consistent to
continue this ground of distinction, and to call everything Art when the
object is to carry out the "doing" (being able), as for example, Art of
building; Science, when merely knowledge is the object; as Science of
mathematics, of astronomy. That in every Art certain complete sciences
may be included is intelligible of itself, and should not perplex us.
But still it is worth observing that there is also no science without a
mixture of Art. In mathematics, for instance, the use of figures and
of algebra is an Art, but that is only one amongst many instances. The
reason is, that however plain and palpable the difference is between
knowledge and power in the composite results of human knowledge, yet it
is difficult to trace out their line of separation in man himself.

2. DIFFICULTY OF SEPARATING PERCEPTION FROM JUDGMENT.

(ART OF WAR.)

All thinking is indeed Art. Where the logician draws the line, where the

premises stop which are the result of cognition--where judgment begins,
there Art begins. But more than this even the perception of the mind is
judgment again, and consequently Art; and at last, even the perception
by the senses as well. In a word, if it is impossible to imagine a human
being possessing merely the faculty of cognition, devoid of judgment or
the reverse, so also Art and Science can never be completely separated
from each other. The more these subtle elements of light embody
themselves in the outward forms of the world, so much the more separate
appear their domains; and now once more, where the object is creation
and production, there is the province of Art; where the object is
investigation and knowledge Science holds sway.--After all this it
results of itself that it is more fitting to say Art of War than Science
of War.

So much for this, because we cannot do without these conceptions. But
now we come forward with the assertion that War is neither an Art nor a
Science in the real signification, and that it is just the setting out
from that starting-point of ideas which has led to a wrong direction
being taken, which has caused War to be put on a par with other arts and
sciences, and has led to a number of erroneous analogies.

This has indeed been felt before now, and on that it was maintained that
War is a handicraft; but there was more lost than gained by that, for
a handicraft is only an inferior art, and as such is also subject to
definite and rigid laws. In reality the Art of War did go on for some
time in the spirit of a handicraft--we allude to the times of the
Condottieri--but then it received that direction, not from intrinsic but
from external causes; and military history shows how little it was at
that time in accordance with the nature of the thing.

3. WAR IS PART OF THE INTERCOURSE OF THE HUMAN RACE.

We say therefore War belongs not to the province of Arts and Sciences,
but to the province of social life. It is a conflict of great interests
which is settled by bloodshed, and only in that is it different from
others. It would be better, instead of comparing it with any Art, to
liken it to business competition, which is also a conflict of human
interests and activities; and it is still more like State policy, which
again, on its part, may be looked upon as a kind of business competition
on a great scale. Besides, State policy is the womb in which War is
developed, in which its outlines lie hidden in a rudimentary state, like
the qualities of living creatures in their germs.(*)

> (*) The analogy has become much closer since Clausewitz's
> time. Now that the first business of the State is regarded
> as the development of facilities for trade, War between
> great nations is only a question of time. No Hague
> Conferences can avert it--EDITOR.

4. DIFFERENCE.

The essential difference consists in this, that War is no activity of
the will, which exerts itself upon inanimate matter like the mechanical
Arts; or upon a living but still passive and yielding subject, like
the human mind and the human feelings in the ideal Arts, but against
a living and reacting force. How little the categories of Arts and
Sciences are applicable to such an activity strikes us at once; and we
can understand at the same time how that constant seeking and striving
after laws like those which may be developed out of the dead material
world could not but lead to constant errors. And yet it is just the
mechanical Arts that some people would imitate in the Art of War. The
imitation of the ideal Arts was quite out of the question, because these
themselves dispense too much with laws and rules, and those hitherto
tried, always acknowledged as insufficient and one-sided, are
perpetually undermined and washed away by the current of opinions,
feelings, and customs.

Page 70

Whether such a conflict of the living, as takes place and is settled in War, is subject to general laws, and whether these are capable of indicating a useful line of action, will be partly investigated in this book; but so much is evident in itself, that this, like every other subject which does not surpass our powers of understanding, may be lighted up, and be made more or less plain in its inner relations by an inquiring mind, and that alone is sufficient to realise the idea of a THEORY.

CHAPTER IV. METHODICISM

IN order to explain ourselves clearly as to the conception of method, and method of action, which play such an important part in War, we must be allowed to cast a hasty glance at the logical hierarchy through which, as through regularly constituted official functionaries, the world of action is governed.

LAW, in the widest sense strictly applying to perception as well as action, has plainly something subjective and arbitrary in its literal meaning, and expresses just that on which we and those things external to us are dependent. As a subject of cognition, LAW is the relation of things and their effects to one another; as a subject of the will, it is a motive of action, and is then equivalent to COMMAND or PROHIBITION.

PRINCIPLE is likewise such a law for action, except that it has not the formal definite meaning, but is only the spirit and sense of law in order to leave the judgment more freedom of application when the diversity of the real world cannot be laid hold of under the definite form of a law. As the judgment must of itself suggest the cases in which the principle is not applicable, the latter therefore becomes in that way a real aid or guiding star for the person acting.

Principle is OBJECTIVE when it is the result of objective truth, and consequently of equal value for all men; it is SUBJECTIVE, and then generally called MAXIM if there are subjective relations in it, and if it therefore has a certain value only for the person himself who makes it.

RULE is frequently taken in the sense of LAW, and then means the same as Principle, for we say "no rule without exceptions," but we do not say "no law without exceptions," a sign that with RULE we retain to ourselves more freedom of application.

In another meaning RULE is the means used of discerning a recondite truth in a particular sign lying close at hand, in order to attach to this particular sign the law of action directed upon the whole truth. Of this kind are all the rules of games of play, all abridged processes in mathematics, &c.

DIRECTIONS and INSTRUCTIONS are determinations of action which have an influence upon a number of minor circumstances too numerous and unimportant for general laws.

Lastly, METHOD, MODE OF ACTING, is an always recurring proceeding selected out of several possible ones; and METHODICISM (METHODISMUS) is that which is determined by methods instead of by general principles or particular prescriptions. By this the cases which are placed under such methods must necessarily be supposed alike in their essential parts. As they cannot all be this, then the point is that at least as many as possible should be; in other words, that Method should be calculated on the most probable cases. Methodicism is therefore not founded on determined particular premises, but on the average probability of cases one with another; and its ultimate tendency is to set up an average truth, the constant and uniform, application of which soon acquires something of the nature of a mechanical appliance, which in the end does that which is right almost unwittingly.

The conception of law in relation to perception is not necessary for the
conduct of War, because the complex phenomena of War are not so regular,
and the regular are not so complex, that we should gain anything more by
this conception than by the simple truth. And where a simple conception
and language is sufficient, to resort to the complex becomes affected
and pedantic. The conception of law in relation to action cannot be used
in the theory of the conduct of War, because owing to the variableness
and diversity of the phenomena there is in it no determination of such a
general nature as to deserve the name of law.

But principles, rules, prescriptions, and methods are conceptions
indispensable to a theory of the conduct of War, in so far as that
theory leads to positive doctrines, because in doctrines the truth can
only crystallise itself in such forms.

As tactics is the branch of the conduct of War in which theory can
attain the nearest to positive doctrine, therefore these conceptions
will appear in it most frequently.

Not to use cavalry against unbroken infantry except in some case of
special emergency, only to use firearms within effective range in
the combat, to spare the forces as much as possible for the final
struggle--these are tactical principles. None of them can be applied
absolutely in every case, but they must always be present to the mind of
the Chief, in order that the benefit of the truth contained in them may
not be lost in cases where that truth can be of advantage.

If from the unusual cooking by an enemy's camp his movement is inferred,
if the intentional exposure of troops in a combat indicates a false
attack, then this way of discerning the truth is called rule, because
from a single visible circumstance that conclusion is drawn which
corresponds with the same.

If it is a rule to attack the enemy with renewed vigour, as soon as he
begins to limber up his artillery in the combat, then on this particular
fact depends a course of action which is aimed at the general situation
of the enemy as inferred from the above fact, namely, that he is about
to give up the fight, that he is commencing to draw off his troops, and
is neither capable of making a serious stand while thus drawing off nor
of making his retreat gradually in good order.

REGULATIONS and METHODS bring preparatory theories into the conduct of
War, in so far as disciplined troops are inoculated with them as active
principles. The whole body of instructions for formations, drill, and
field service are regulations and methods: in the drill instructions
the first predominate, in the field service instructions the latter.
To these things the real conduct of War attaches itself; it takes them
over, therefore, as given modes of proceeding, and as such they must
appear in the theory of the conduct of War.

But for those activities retaining freedom in the employment of these
forces there cannot be regulations, that is, definite instructions,
because they would do away with freedom of action. Methods, on the other
hand, as a general way of executing duties as they arise, calculated, as
we have said, on an average of probability, or as a dominating influence
of principles and rules carried through to application, may certainly
appear in the theory of the conduct of War, provided only they are
not represented as something different from what they are, not as the
absolute and necessary modes of action (systems), but as the best of
general forms which may be used as shorter ways in place of a particular
disposition for the occasion, at discretion.

But the frequent application of methods will be seen to be most
essential and unavoidable in the conduct of War, if we reflect how much
action proceeds on mere conjecture, or in complete uncertainty,
because one side is prevented from learning all the circumstances which

On War.txt
influence the dispositions of the other, or because, even if these
circumstances which influence the decisions of the one were really
known, there is not, owing to their extent and the dispositions they
would entail, sufficient time for the other to carry out all necessary
counteracting measures--that therefore measures in War must always
be calculated on a certain number of possibilities; if we reflect how
numberless are the trifling things belonging to any single event, and
which therefore should be taken into account along with it, and that
therefore there is no other means to suppose the one counteracted by
the other, and to base our arrangements only upon what is of a general
nature and probable; if we reflect lastly that, owing to the increasing
number of officers as we descend the scale of rank, less must be left
to the true discernment and ripe judgment of individuals the lower the
sphere of action, and that when we reach those ranks where we can look
for no other notions but those which the regulations of the service and
experience afford, we must help them with the methodic forms bordering
on those regulations. This will serve both as a support to their
judgment and a barrier against those extravagant and erroneous views
which are so especially to be dreaded in a sphere where experience is so
costly.

Besides this absolute need of method in action, we must also acknowledge
that it has a positive advantage, which is that, through the constant
repetition of a formal exercise, a readiness, precision, and firmness
is attained in the movement of troops which diminishes the natural
friction, and makes the machine move easier.

Method will therefore be the more generally used, become the more
indispensable, the farther down the scale of rank the position of the
active agent; and on the other hand, its use will diminish upwards,
until in the highest position it quite disappears. For this reason it is
more in its place in tactics than in strategy.

War in its highest aspects consists not of an infinite number of little
events, the diversities in which compensate each other, and which
therefore by a better or worse method are better or worse governed, but
of separate great decisive events which must be dealt with separately.
It is not like a field of stalks, which, without any regard to the
particular form of each stalk, will be mowed better or worse, according
as the mowing instrument is good or bad, but rather as a group of large
trees, to which the axe must be laid with judgment, according to the
particular form and inclination of each separate trunk.

How high up in military activity the admissibility of method in action
reaches naturally determines itself, not according to actual rank, but
according to things; and it affects the highest positions in a less
degree, only because these positions have the most comprehensive
subjects of activity. A constant order of battle, a constant formation
of advance guards and outposts, are methods by which a General ties
not only his subordinates' hands, but also his own in certain cases.
Certainly they may have been devised by himself, and may be applied
by him according to circumstances, but they may also be a subject of
theory, in so far as they are based on the general properties of troops
and weapons. On the other hand, any method by which definite plans
for wars or campaigns are to be given out all ready made as if from a
machine are absolutely worthless.

As long as there exists no theory which can be sustained, that is, no
enlightened treatise on the conduct of War, method in action cannot but
encroach beyond its proper limits in high places, for men employed
in these spheres of activity have not always had the opportunity of
educating themselves, through study and through contact with the higher
interests. In the impracticable and inconsistent disquisitions of
theorists and critics they cannot find their way, their sound common
sense rejects them, and as they bring with them no knowledge but that
derived from experience, therefore in those cases which admit of, and
require, a free individual treatment they readily make use of the means
which experience gives them--that is, an imitation of the particular
Page 73

On War.txt
methods practised by great Generals, by which a method of action then
arises of itself. If we see Frederick the Great's Generals always making
their appearance in the so-called oblique order of battle, the Generals
of the French Revolution always using turning movements with a long,
extended line of battle, and Buonaparte's lieutenants rushing to the
attack with the bloody energy of concentrated masses, then we recognise
in the recurrence of the mode of proceeding evidently an adopted
method, and see therefore that method of action can reach up to regions
bordering on the highest. Should an improved theory facilitate the study
of the conduct of War, form the mind and judgment of men who are rising
to the highest commands, then also method in action will no longer reach
so far, and so much of it as is to be considered indispensable will then
at least be formed from theory itself, and not take place out of mere
imitation. However pre-eminently a great Commander does things, there
is always something subjective in the way he does them; and if he has
a certain manner, a large share of his individuality is contained in it
which does not always accord with the individuality of the person who
copies his manner.

At the same time, it would neither be possible nor right to banish
subjective methodicism or manner completely from the conduct of War: it
is rather to be regarded as a manifestation of that influence which the
general character of a War has upon its separate events, and to which
satisfaction can only be done in that way if theory is not able to
foresee this general character and include it in its considerations.
What is more natural than that the War of the French Revolution had its
own way of doing things? and what theory could ever have included that
peculiar method? The evil is only that such a manner originating in
a special case easily outlives itself, because it continues whilst
circumstances imperceptibly change. This is what theory should prevent
by lucid and rational criticism. When in the year 1806 the Prussian
Generals, Prince Louis at Saalfeld, Tauentzien on the Dornberg near
Jena, Grawert before and Ruechel behind Kappellendorf, all threw
themselves into the open jaws of destruction in the oblique order of
Frederick the Great, and managed to ruin Hohenlohe's Army in a way that
no Army was ever ruined, even on the field of battle, all this was done
through a manner which had outlived its day, together with the most
downright stupidity to which methodicism ever led.

CHAPTER V. CRITICISM

THE influence of theoretical principles upon real life is produced
more through criticism than through doctrine, for as criticism is an
application of abstract truth to real events, therefore it not only
brings truth of this description nearer to life, but also accustoms the
understanding more to such truths by the constant repetition of their
application. We therefore think it necessary to fix the point of view
for criticism next to that for theory.

From the simple narration of an historical occurrence which places
events in chronological order, or at most only touches on their more
immediate causes, we separate the CRITICAL.

In this CRITICAL three different operations of the mind may be observed.

First, the historical investigation and determining of doubtful facts.
This is properly historical research, and has nothing in common with
theory.

Secondly, the tracing of effects to causes. This is the REAL CRITICAL
INQUIRY; it is indispensable to theory, for everything which in theory
is to be established, supported, or even merely explained, by experience
can only be settled in this way.

Thirdly, the testing of the means employed. This is criticism, properly
speaking, in which praise and censure is contained. This is where theory

On War.txt
helps history, or rather, the teaching to be derived from it.

In these two last strictly critical parts of historical study, all
depends on tracing things to their primary elements, that is to say,
up to undoubted truths, and not, as is so often done, resting half-way,
that is, on some arbitrary assumption or supposition.

As respects the tracing of effect to cause, that is often attended with
the insuperable difficulty that the real causes are not known. In none
of the relations of life does this so frequently happen as in War, where
events are seldom fully known, and still less motives, as the latter
have been, perhaps purposely, concealed by the chief actor, or have been
of such a transient and accidental character that they have been lost
for history. For this reason critical narration must generally proceed
hand in hand with historical investigation, and still such a want of
connection between cause and effect will often present itself, that it
does not seem justifiable to consider effects as the necessary results
of known causes. Here, therefore must occur, that is, historical results
which cannot be made use of for teaching. All that theory can demand is
that the investigation should be rigidly conducted up to that point, and
there leave off without drawing conclusions. A real evil springs up only
if the known is made perforce to suffice as an explanation of effects,
and thus a false importance is ascribed to it.

Besides this difficulty, critical inquiry also meets with another great
and intrinsic one, which is that the progress of events in War seldom
proceeds from one simple cause, but from several in common, and that
it therefore is not sufficient to follow up a series of events to
their origin in a candid and impartial spirit, but that it is then also
necessary to apportion to each contributing cause its due weight. This
leads, therefore, to a closer investigation of their nature, and thus a
critical investigation may lead into what is the proper field of theory.

The critical CONSIDERATION, that is, the testing of the means, leads to
the question, which are the effects peculiar to the means applied,
and whether these effects were comprehended in the plans of the person
directing?

The effects peculiar to the means lead to the investigation of their
nature, and thus again into the field of theory.

We have already seen that in criticism all depends upon attaining
to positive truth; therefore, that we must not stop at arbitrary
propositions which are not allowed by others, and to which other perhaps
equally arbitrary assertions may again be opposed, so that there is no
end to pros and cons; the whole is without result, and therefore without
instruction.

We have seen that both the search for causes and the examination
of means lead into the field of theory; that is, into the field of
universal truth, which does not proceed solely from the case immediately
under examination. If there is a theory which can be used, then the
critical consideration will appeal to the proofs there afforded, and the
examination may there stop. But where no such theoretical truth is to be
found, the inquiry must be pushed up to the original elements. If this
necessity occurs often, it must lead the historian (according to a
common expression) into a labyrinth of details. He then has his hands
full, and it is impossible for him to stop to give the requisite
attention everywhere; the consequence is, that in order to set bounds to
his investigation, he adopts some arbitrary assumptions which, if they
do not appear so to him, do so to others, as they are not evident in
themselves or capable of proof.

A sound theory is therefore an essential foundation for criticism, and
it is impossible for it, without the assistance of a sensible theory,
to attain to that point at which it commences chiefly to be instructive,
that is, where it becomes demonstration, both convincing and sans
re'plique.

But it would be a visionary hope to believe in the possibility of a
theory applicable to every abstract truth, leaving nothing for criticism
to do but to place the case under its appropriate law: it would be
ridiculous pedantry to lay down as a rule for criticism that it must
always halt and turn round on reaching the boundaries of sacred theory.
The same spirit of analytical inquiry which is the origin of theory must
also guide the critic in his work; and it can and must therefore happen
that he strays beyond the boundaries of the province of theory and
elucidates those points with which he is more particularly concerned. It
is more likely, on the contrary, that criticism would completely fail
in its object if it degenerated into a mechanical application of theory.
All positive results of theoretical inquiry, all principles, rules, and
methods, are the more wanting in generality and positive truth the more
they become positive doctrine. They exist to offer themselves for use as
required, and it must always be left for judgment to decide whether
they are suitable or not. Such results of theory must never be used in
criticism as rules or norms for a standard, but in the same way as the
person acting should use them, that is, merely as aids to judgment. If
it is an acknowledged principle in tactics that in the usual order of
battle cavalry should be placed behind infantry, not in line with it,
still it would be folly on this account to condemn every deviation from
this principle. Criticism must investigate the grounds of the deviation,
and it is only in case these are insufficient that it has a right to
appeal to principles laid down in theory. If it is further established
in theory that a divided attack diminishes the probability of success,
still it would be just as unreasonable, whenever there is a divided
attack and an unsuccessful issue, to regard the latter as the result of
the former, without further investigation into the connection between
the two, as where a divided attack is successful to infer from it the
fallacy of that theoretical principle. The spirit of investigation which
belongs to criticism cannot allow either. Criticism therefore supports
itself chiefly on the results of the analytical investigation of theory;
what has been made out and determined by theory does not require to be
demonstrated over again by criticism, and it is so determined by theory
that criticism may find it ready demonstrated.

This office of criticism, of examining the effect produced by certain
causes, and whether a means applied has answered its object, will be
easy enough if cause and effect, means and end, are all near together.

If an Army is surprised, and therefore cannot make a regular and
intelligent use of its powers and resources, then the effect of the
surprise is not doubtful.--If theory has determined that in a battle
the convergent form of attack is calculated to produce greater but
less certain results, then the question is whether he who employs that
convergent form had in view chiefly that greatness of result as his
object; if so, the proper means were chosen. But if by this form he
intended to make the result more certain, and that expectation was
founded not on some exceptional circumstances (in this case), but on the
general nature of the convergent form, as has happened a hundred times,
then he mistook the nature of the means and committed an error.

Here the work of military investigation and criticism is easy, and it
will always be so when confined to the immediate effects and objects.
This can be done quite at option, if we abstract the connection of the
parts with the whole, and only look at things in that relation.

But in War, as generally in the world, there is a connection between
everything which belongs to a whole; and therefore, however small a
cause may be in itself, its effects reach to the end of the act of
warfare, and modify or influence the final result in some degree, let
that degree be ever so small. In the same manner every means must be
felt up to the ultimate object.

We can therefore trace the effects of a cause as long as events are
worth noticing, and in the same way we must not stop at the testing of a
means for the immediate object, but test also this object as a means to

a higher one, and thus ascend the series of facts in succession, until
we come to one so absolutely necessary in its nature as to require no
examination or proof. In many cases, particularly in what concerns great
and decisive measures, the investigation must be carried to the final
aim, to that which leads immediately to peace.

It is evident that in thus ascending, at every new station which we
reach a new point of view for the judgment is attained, so that the same
means which appeared advisable at one station, when looked at from the
next above it may have to be rejected.

The search for the causes of events and the comparison of means with
ends must always go hand in hand in the critical review of an act, for
the investigation of causes leads us first to the discovery of those
things which are worth examining.

This following of the clue up and down is attended with considerable
difficulty, for the farther from an event the cause lies which we are
looking for, the greater must be the number of other causes which must
at the same time be kept in view and allowed for in reference to the
share which they have in the course of events, and then eliminated,
because the higher the importance of a fact the greater will be the
number of separate forces and circumstances by which it is conditioned.
If we have unravelled the causes of a battle being lost, we have
certainly also ascertained a part of the causes of the consequences
which this defeat has upon the whole War, but only a part, because the
effects of other causes, more or less according to circumstances, will
flow into the final result.

The same multiplicity of circumstances is presented also in the
examination of the means the higher our point of view, for the higher
the object is situated, the greater must be the number of means employed
to reach it. The ultimate object of the War is the object aimed at by
all the Armies simultaneously, and it is therefore necessary that the
consideration should embrace all that each has done or could have done.

It is obvious that this may sometimes lead to a wide field of inquiry,
in which it is easy to wander and lose the way, and in which this
difficulty prevails--that a number of assumptions or suppositions must
be made about a variety of things which do not actually appear, but
which in all probability did take place, and therefore cannot possibly
be left out of consideration.

When Buonaparte, in 1797,(*) at the head of the Army of Italy, advanced
from the Tagliamento against the Archduke Charles, he did so with a view
to force that General to a decisive action before the reinforcements
expected from the Rhine had reached him. If we look, only at the
immediate object, the means were well chosen and justified by the
result, for the Archduke was so inferior in numbers that he only made a
show of resistance on the Tagliamento, and when he saw his adversary so
strong and resolute, yielded ground, and left open the passages, of
the Norican Alps. Now to what use could Buonaparte turn this fortunate
event? To penetrate into the heart of the Austrian empire itself, to
facilitate the advance of the Rhine Armies under Moreau and Hoche, and
open communication with them? This was the view taken by Buonaparte,
and from this point of view he was right. But now, if criticism places
itself at a higher point of view--namely, that of the French Directory,
which body could see and know that the Armies on the Rhine could not
commence the campaign for six weeks, then the advance of Buonaparte over
the Norican Alps can only be regarded as an extremely hazardous
measure; for if the Austrians had drawn largely on their Rhine Armies
to reinforce their Army in Styria, so as to enable the Archduke to fall
upon the Army of Italy, not only would that Army have been routed, but
the whole campaign lost. This consideration, which attracted the serious
attention of Buonaparte at Villach, no doubt induced him to sign the
armistice of Leoben with so much readiness.

(*) Compare Hinterlassene Werke, 2nd edition, vol. iv. p.

If criticism takes a still higher position, and if it knows that the
Austrians had no reserves between the Army of the Archduke Charles and
Vienna, then we see that Vienna became threatened by the advance of the
Army of Italy.

Supposing that Buonaparte knew that the capital was thus uncovered, and
knew that he still retained the same superiority in numbers over the
Archduke as he had in Styria, then his advance against the heart of the
Austrian States was no longer without purpose, and its value depended on
the value which the Austrians might place on preserving their capital.
If that was so great that, rather than lose it, they would accept the
conditions of peace which Buonaparte was ready to offer them, it became
an object of the first importance to threaten Vienna. If Buonaparte
had any reason to know this, then criticism may stop there, but if this
point was only problematical, then criticism must take a still higher
position, and ask what would have followed if the Austrians had resolved
to abandon Vienna and retire farther into the vast dominions still left
to them. But it is easy to see that this question cannot be answered
without bringing into the consideration the probable movements of the
Rhine Armies on both sides. Through the decided superiority of numbers
on the side of the French--130,000 to 80,000--there could be little
doubt of the result; but then next arises the question, What use would
the Directory make of a victory; whether they would follow up their
success to the opposite frontiers of the Austrian monarchy, therefore
to the complete breaking up or overthrow of that power, or whether they
would be satisfied with the conquest of a considerable portion to
serve as a security for peace? The probable result in each case must
be estimated, in order to come to a conclusion as to the probable
determination of the Directory. Supposing the result of these
considerations to be that the French forces were much too weak for the
complete subjugation of the Austrian monarchy, so that the attempt might
completely reverse the respective positions of the contending Armies,
and that even the conquest and occupation of a considerable district of
country would place the French Army in strategic relations to which they
were not equal, then that result must naturally influence the estimate
of the position of the Army of Italy, and compel it to lower its
expectations. And this, it was no doubt which influenced Buonaparte,
although fully aware of the helpless condition of the Archduke, still to
sign the peace of Campo Formio, which imposed no greater sacrifices on
the Austrians than the loss of provinces which, even if the campaign
took the most favourable turn for them, they could not have reconquered.
But the French could not have reckoned on even the moderate treaty
of Campo Formio, and therefore it could not have been their object
in making their bold advance if two considerations had not presented
themselves to their view, the first of which consisted in the question,
what degree of value the Austrians would attach to each of the
above-mentioned results; whether, notwithstanding the probability of a
satisfactory result in either of these cases, would it be worth while to
make the sacrifices inseparable from a continuance of the War, when
they could be spared those sacrifices by a peace on terms not too
humiliating? The second consideration is the question whether the
Austrian Government, instead of seriously weighing the possible results
of a resistance pushed to extremities, would not prove completely
disheartened by the impression of their present reverses.

The consideration which forms the subject of the first is no idle piece
of subtle argument, but a consideration of such decidedly practical
importance that it comes up whenever the plan of pushing War to the
utmost extremity is mooted, and by its weight in most cases restrains
the execution of such plans.

The second consideration is of equal importance, for we do not make War
with an abstraction but with a reality, which we must always keep
in view, and we may be sure that it was not overlooked by the bold
Buonaparte--that is, that he was keenly alive to the terror which the
appearance of his sword inspired. It was reliance on that which led him

to Moscow. There it led him into a scrape. The terror of him had been
weakened by the gigantic struggles in which he had been engaged; in the
year 1797 it was still fresh, and the secret of a resistance pushed
to extremities had not been discovered; nevertheless even in 1797 his
boldness might have led to a negative result if, as already said, he had
not with a sort of presentiment avoided it by signing the moderate peace
of Campo Formio.

We must now bring these considerations to a close--they will suffice
to show the wide sphere, the diversity and embarrassing nature of the
subjects embraced in a critical examination carried to the fullest
extent, that is, to those measures of a great and decisive class which
must necessarily be included. It follows from them that besides a
theoretical acquaintance with the subject, natural talent must also have
a great influence on the value of critical examinations, for it
rests chiefly with the latter to throw the requisite light on the
interrelations of things, and to distinguish from amongst the endless
connections of events those which are really essential.

But talent is also called into requisition in another way. Critical
examination is not merely the appreciation of those means which have
been actually employed, but also of all possible means, which therefore
must be suggested in the first place--that is, must be discovered; and
the use of any particular means is not fairly open to censure until
a better is pointed out. Now, however small the number of possible
combinations may be in most cases, still it must be admitted that to
point out those which have not been used is not a mere analysis of
actual things, but a spontaneous creation which cannot be prescribed,
and depends on the fertility of genius.

We are far from seeing a field for great genius in a case which admits
only of the application of a few simple combinations, and we think it
exceedingly ridiculous to hold up, as is often done, the turning of a
position as an invention showing the highest genius; still nevertheless
this creative self-activity on the part of the critic is necessary,
and it is one of the points which essentially determine the value of
critical examination.

When Buonaparte on 30th July, 1796,(*) determined to raise the siege
of Mantua, in order to march with his whole force against the enemy,
advancing in separate columns to the relief of the place, and to beat
them in detail, this appeared the surest way to the attainment of
brilliant victories. These victories actually followed, and were
afterwards again repeated on a still more brilliant scale on the attempt
to relieve the fortress being again renewed. We hear only one opinion on
these achievements, that of unmixed admiration.

(*) Compare Hinterlassene Werke, 2nd edition, vol. iv. p.
107 et seq.

At the same time, Buonaparte could not have adopted this course on
the 30th July without quite giving up the idea of the siege of Mantua,
because it was impossible to save the siege train, and it could not be
replaced by another in this campaign. In fact, the siege was converted
into a blockade, and the town, which if the siege had continued
must have very shortly fallen, held out for six months in spite of
Buonaparte's victories in the open field.

Criticism has generally regarded this as an evil that was unavoidable,
because critics have not been able to suggest any better course.
Resistance to a relieving Army within lines of circumvallation had
fallen into such disrepute and contempt that it appears to have entirely
escaped consideration as a means. And yet in the reign of Louis XIV.
that measure was so often used with success that we can only attribute
to the force of fashion the fact that a hundred years later it
never occurred to any one even to propose such a measure. If the
practicability of such a plan had ever been entertained for a moment,
a closer consideration of circumstances would have shown that 40,000 of

the best infantry in the world under Buonaparte, behind strong lines of
circumvallation round Mantua, had so little to fear from the 50,000 men
coming to the relief under Wurmser, that it was very unlikely that any
attempt even would be made upon their lines. We shall not seek here to
establish this point, but we believe enough has been said to show
that this means was one which had a right to a share of consideration.
Whether Buonaparte himself ever thought of such a plan we leave
undecided; neither in his memoirs nor in other sources is there any
trace to be found of his having done so; in no critical works has it
been touched upon, the measure being one which the mind had lost sight
of. The merit of resuscitating the idea of this means is not great, for
it suggests itself at once to any one who breaks loose from the trammels
of fashion. Still it is necessary that it should suggest itself for
us to bring it into consideration and compare it with the means which
Buonaparte employed. Whatever may be the result of the comparison, it is
one which should not be omitted by criticism.

When Buonaparte, in February, 1814,(*) after gaining the battles at
Etoges, Champ-Aubert, and Montmirail, left Bluecher's Army, and turning
upon Schwartzenberg, beat his troops at Montereau and Mormant, every
one was filled with admiration, because Buonaparte, by thus throwing his
concentrated force first upon one opponent, then upon another, made a
brilliant use of the mistakes which his adversaries had committed
in dividing their forces. If these brilliant strokes in different
directions failed to save him, it was generally considered to be no
fault of his, at least. No one has yet asked the question, What
would have been the result if, instead of turning from Bluecher upon
Schwartzenberg, he had tried another blow at Bluecher, and pursued him
to the Rhine? We are convinced that it would have completely changed
the course of the campaign, and that the Army of the Allies, instead of
marching to Paris, would have retired behind the Rhine. We do not ask
others to share our conviction, but no one who understands the thing
will doubt, at the mere mention of this alternative course, that it is
one which should not be overlooked in criticism.

(*) Compare Hinterlassene Werks, 2nd edition. vol. vii. p.
193 et seq.

In this case the means of comparison lie much more on the surface
than in the foregoing, but they have been equally overlooked, because
one-sided views have prevailed, and there has been no freedom of
judgment.

From the necessity of pointing out a better means which might have
been used in place of those which are condemned has arisen the form of
criticism almost exclusively in use, which contents itself with pointing
out the better means without demonstrating in what the superiority
consists. The consequence is that some are not convinced, that others
start up and do the same thing, and that thus discussion arises which
is without any fixed basis for the argument. Military literature abounds
with matter of this sort.

The demonstration we require is always necessary when the superiority
of the means propounded is not so evident as to leave no room for doubt,
and it consists in the examination of each of the means on its own
merits, and then of its comparison with the object desired. When once
the thing is traced back to a simple truth, controversy must cease, or
at all events a new result is obtained, whilst by the other plan the
pros and cons go on for ever consuming each other.

Should we, for example, not rest content with assertion in the case
before mentioned, and wish to prove that the persistent pursuit
of Bluecher would have been more advantageous than the turning on
Schwartzenberg, we should support the arguments on the following simple
truths:

1. In general it is more advantageous to continue our blows in one
and the same direction, because there is a loss of time in striking in

On War.txt
different directions; and at a point where the moral power is already
shaken by considerable losses there is the more reason to expect fresh
successes, therefore in that way no part of the preponderance already
gained is left idle.

2. Because Bluecher, although weaker than Schwartzenberg, was, on
account of his enterprising spirit, the more important adversary; in
him, therefore, lay the centre of attraction which drew the others along
in the same direction.

3. Because the losses which Bluecher had sustained almost amounted to a
defeat, which gave Buonaparte such a preponderance over him as to
make his retreat to the Rhine almost certain, and at the same time no
reserves of any consequence awaited him there.

4. Because there was no other result which would be so terrific in its
aspects, would appear to the imagination in such gigantic proportions,
an immense advantage in dealing with a Staff so weak and irresolute as
that of Schwartzenberg notoriously was at this time. What had
happened to the Crown Prince of Wartemberg at Montereau, and to Count
Wittgenstein at Mormant, Prince Schwartzenberg must have known well
enough; but all the untoward events on Bluecher's distant and separate
line from the Marne to the Rhine would only reach him by the avalanche
of rumour. The desperate movements which Buonaparte made upon Vitry at
the end of March, to see what the Allies would do if he threatened to
turn them strategically, were evidently done on the principle of working
on their fears; but it was done under far different circumstances, in
consequence of his defeat at Laon and Arcis, and because Bluecher, with
100,000 men, was then in communication with Schwartzenberg.

There are people, no doubt, who will not be convinced on these
arguments, but at all events they cannot retort by saying, that "whilst
Buonaparte threatened Schwartzenberg's base by advancing to the Rhine,
Schwartzenberg at the same time threatened Buonaparte's communications
with Paris," because we have shown by the reasons above given that
Schwartzenberg would never have thought of marching on Paris.

With respect to the example quoted by us from the campaign of 1796, we
should say: Buonaparte looked upon the plan he adopted as the surest
means of beating the Austrians; but admitting that it was so, still the
object to be attained was only an empty victory, which could have hardly
any sensible influence on the fall of Mantua. The way which we should
have chosen would, in our opinion, have been much more certain to
prevent the relief of Mantua; but even if we place ourselves in the
position of the French General and assume that it was not so, and look
upon the certainty of success to have been less, the question then
amounts to a choice between a more certain but less useful, and
therefore less important, victory on the one hand, and a somewhat less
probable but far more decisive and important victory, on the other
hand. Presented in this form, boldness must have declared for the second
solution, which is the reverse of what took place, when the thing
was only superficially viewed. Buonaparte certainly was anything but
deficient in boldness, and we may be sure that he did not see the whole
case and its consequences as fully and clearly as we can at the present
time.

Naturally the critic, in treating of the means, must often appeal to
military history, as experience is of more value in the Art of War
than all philosophical truth. But this exemplification from history
is subject to certain conditions, of which we shall treat in a special
chapter and unfortunately these conditions are so seldom regarded that
reference to history generally only serves to increase the confusion of
ideas.

We have still a most important subject to consider, which is, How far
criticism in passing judgments on particular events is permitted, or in
duty bound, to make use of its wider view of things, and therefore also
of that which is shown by results; or when and where it should leave out
Page 81

of sight these things in order to place itself, as far as possible, in
the exact position of the chief actor?

If criticism dispenses praise or censure, it should seek to place itself
as nearly as possible at the same point of view as the person acting,
that is to say, to collect all he knew and all the motives on which he
acted, and, on the other hand, to leave out of the consideration all
that the person acting could not or did not know, and above all, the
result. But this is only an object to aim at, which can never be reached
because the state of circumstances from which an event proceeded can
never be placed before the eye of the critic exactly as it lay before
the eye of the person acting. A number of inferior circumstances, which
must have influenced the result, are completely lost to sight, and many
a subjective motive has never come to light.

The latter can only be learnt from the memoirs of the chief actor, or
from his intimate friends; and in such things of this kind are often
treated of in a very desultory manner, or purposely misrepresented.
Criticism must, therefore, always forego much which was present in the
minds of those whose acts are criticised.

On the other hand, it is much more difficult to leave out of sight that
which criticism knows in excess. This is only easy as regards accidental
circumstances, that is, circumstances which have been mixed up, but are
in no way necessarily related. But it is very difficult, and, in fact,
can never be completely done with regard to things really essential.

Let us take first, the result. If it has not proceeded from accidental
circumstances, it is almost impossible that the knowledge of it should
not have an effect on the judgment passed on events which have preceded
it, for we see these things in the light of this result, and it is to
a certain extent by it that we first become acquainted with them and
appreciate them. Military history, with all its events, is a source of
instruction for criticism itself, and it is only natural that criticism
should throw that light on things which it has itself obtained from the
consideration of the whole. If therefore it might wish in some cases to
leave the result out of the consideration, it would be impossible to do
so completely.

But it is not only in relation to the result, that is, with what takes
place at the last, that this embarrassment arises; the same occurs in
relation to preceding events, therefore with the data which furnished
the motives to action. Criticism has before it, in most cases, more
information on this point than the principal in the transaction. Now
it may seem easy to dismiss from the consideration everything of
this nature, but it is not so easy as we may think. The knowledge
of preceding and concurrent events is founded not only on certain
information, but on a number of conjectures and suppositions; indeed,
there is hardly any of the information respecting things not purely
accidental which has not been preceded by suppositions or conjectures
destined to take the place of certain information in case such should
never be supplied. Now is it conceivable that criticism in after
times, which has before it as facts all the preceding and concurrent
circumstances, should not allow itself to be thereby influenced when it
asks itself the question, what portion of the circumstances, which at
the moment of action were unknown, would it have held to be probable? We
maintain that in this case, as in the case of the results, and for the
same reason, it is impossible to disregard all these things completely.

If therefore the critic wishes to bestow praise or blame upon any single
act, he can only succeed to a certain degree in placing himself in the
position of the person whose act he has under review. In many cases
he can do so sufficiently near for any practical purpose, but in
many instances it is the very reverse, and this fact should never be
overlooked.

But it is neither necessary nor desirable that criticism should
completely identify itself with the person acting. In War, as in all

matters of skill, there is a certain natural aptitude required which
is called talent. This may be great or small. In the first case it may
easily be superior to that of the critic, for what critic can pretend to
the skill of a Frederick or a Buonaparte? Therefore, if criticism is not
to abstain altogether from offering an opinion where eminent talent is
concerned, it must be allowed to make use of the advantage which its
enlarged horizon affords. Criticism must not, therefore, treat the
solution of a problem by a great General like a sum in arithmetic; it
is only through the results and through the exact coincidences of events
that it can recognise with admiration how much is due to the exercise
of genius, and that it first learns the essential combination which the
glance of that genius devised.

But for every, even the smallest, act of genius it is necessary that
criticism should take a higher point of view, so that, having at command
many objective grounds of decision, it may be as little subjective as
possible, and that the critic may not take the limited scope of his own
mind as a standard.

This elevated position of criticism, its praise and blame pronounced
with a full knowledge of all the circumstances, has in itself nothing
which hurts our feelings; it only does so if the critic pushes himself
forward, and speaks in a tone as if all the wisdom which he has obtained
by an exhaustive examination of the event under consideration were
really his own talent. Palpable as is this deception, it is one which
people may easily fall into through vanity, and one which is naturally
distasteful to others. It very often happens that although the critic
has no such arrogant pretensions, they are imputed to him by the
reader because he has not expressly disclaimed them, and then follows
immediately a charge of a want of the power of critical judgment.

If therefore a critic points out an error made by a Frederick or a
Buonaparte, that does not mean that he who makes the criticism would not
have committed the same error; he may even be ready to grant that had
he been in the place of these great Generals he might have made much
greater mistakes; he merely sees this error from the chain of events,
and he thinks that it should not have escaped the sagacity of the
General.

This is, therefore, an opinion formed through the connection of events,
and therefore through the RESULT. But there is another quite different
effect of the result itself upon the judgment, that is if it is used
quite alone as an example for or against the soundness of a measure.
This may be called JUDGMENT ACCORDING TO THE RESULT. Such a judgment
appears at first sight inadmissible, and yet it is not.

When Buonaparte marched to Moscow in 1812, all depended upon whether the
taking of the capital, and the events which preceded the capture, would
force the Emperor Alexander to make peace, as he had been compelled to
do after the battle of Friedland in 1807, and the Emperor Francis in
1805 and 1809 after Austerlitz and Wagram; for if Buonaparte did not
obtain a peace at Moscow, there was no alternative but to return--that
is, there was nothing for him but a strategic defeat. We shall leave out
of the question what he did to get to Moscow, and whether in his advance
he did not miss many opportunities of bringing the Emperor Alexander
to peace; we shall also exclude all consideration of the disastrous
circumstances which attended his retreat, and which perhaps had their
origin in the general conduct of the campaign. Still the question
remains the same, for however much more brilliant the course of the
campaign up to Moscow might have been, still there was always an
uncertainty whether the Emperor Alexander would be intimidated into
making peace; and then, even if a retreat did not contain in itself the
seeds of such disasters as did in fact occur, still it could never be
anything else than a great strategic defeat. If the Emperor Alexander
agreed to a peace which was disadvantageous to him, the campaign of 1812
would have ranked with those of Austerlitz, Friedland, and Wagram.
But these campaigns also, if they had not led to peace, would in all
probability have ended in similar catastrophes. Whatever, therefore,

of genius, skill, and energy the Conqueror of the World applied to the
task, this last question addressed to fate(*) remained always the same.
Shall we then discard the campaigns of 1805, 1807, 1809, and say on
account of the campaign of 1812 that they were acts of imprudence;
that the results were against the nature of things, and that in 1812
strategic justice at last found vent for itself in opposition to blind
chance? That would be an unwarrantable conclusion, a most arbitrary
judgment, a case only half proved, because no human, eye can trace the
thread of the necessary connection of events up to the determination of
the conquered Princes.

(*) "Frage an der Schicksal,"a familiar quotation from
Schiller.--TR.

Still less can we say the campaign of 1812 merited the same success
as the others, and that the reason why it turned out otherwise lies in
something unnatural, for we cannot regard the firmness of Alexander as
something unpredictable.

What can be more natural than to say that in the years 1805, 1807, 1809,
Buonaparte judged his opponents correctly, and that in 1812 he erred
in that point? On the former occasions, therefore, he was right, in the
latter wrong, and in both cases we judge by the RESULT.

All action in War, as we have already said, is directed on probable,
not on certain, results. Whatever is wanting in certainty must always be
left to fate, or chance, call it which you will. We may demand that what
is so left should be as little as possible, but only in relation to the
particular case--that is, as little as is possible in this one case, but
not that the case in which the least is left to chance is always to
be preferred. That would be an enormous error, as follows from all our
theoretical views. There are cases in which the greatest daring is the
greatest wisdom.

Now in everything which is left to chance by the chief actor, his
personal merit, and therefore his responsibility as well, seems to be
completely set aside; nevertheless we cannot suppress an inward
feeling of satisfaction whenever expectation realises itself, and if it
disappoints us our mind is dissatisfied; and more than this of right and
wrong should not be meant by the judgment which we form from the mere
result, or rather that we find there.

Nevertheless, it cannot be denied that the satisfaction which our mind
experiences at success, the pain caused by failure, proceed from a sort
of mysterious feeling; we suppose between that success ascribed to good
fortune and the genius of the chief a fine connecting thread, invisible
to the mind's eye, and the supposition gives pleasure. What tends to
confirm this idea is that our sympathy increases, becomes more decided,
if the successes and defeats of the principal actor are often repeated.
Thus it becomes intelligible how good luck in War assumes a much nobler
nature than good luck at play. In general, when a fortunate warrior does
not otherwise lessen our interest in his behalf, we have a pleasure in
accompanying him in his career.

Criticism, therefore, after having weighed all that comes within the
sphere of human reason and conviction, will let the result speak for
that part where the deep mysterious relations are not disclosed in
any visible form, and will protect this silent sentence of a higher
authority from the noise of crude opinions on the one hand, while on
the other it prevents the gross abuse which might be made of this last
tribunal.

This verdict of the result must therefore always bring forth that which
human sagacity cannot discover; and it will be chiefly as regards
the intellectual powers and operations that it will be called into
requisition, partly because they can be estimated with the least
certainty, partly because their close connection with the will is
favourable to their exercising over it an important influence. When

fear or bravery precipitates the decision, there is nothing objective
intervening between them for our consideration, and consequently nothing
by which sagacity and calculation might have met the probable result.

We must now be allowed to make a few observations on the instrument of
criticism, that is, the language which it uses, because that is to
a certain extent connected with the action in War; for the critical
examination is nothing more than the deliberation which should precede
action in War. We therefore think it very essential that the language
used in criticism should have the same character as that which
deliberation in War must have, for otherwise it would cease to be
practical, and criticism could gain no admittance in actual life.

We have said in our observations on the theory of the conduct of War
that it should educate the mind of the Commander for War, or that its
teaching should guide his education; also that it is not intended to
furnish him with positive doctrines and systems which he can use like
mental appliances. But if the construction of scientific formulae is
never required, or even allowable, in War to aid the decision on the
case presented, if truth does not appear there in a systematic shape,
if it is not found in an indirect way, but directly by the natural
perception of the mind, then it must be the same also in a critical
review.

It is true as we have seen that, wherever complete demonstration of the
nature of things would be too tedious, criticism must support itself on
those truths which theory has established on the point. But, just as in
War the actor obeys these theoretical truths rather because his mind is
imbued with them than because he regards them as objective inflexible
laws, so criticism must also make use of them, not as an external law
or an algebraic formula, of which fresh proof is not required each time
they are applied, but it must always throw a light on this proof itself,
leaving only to theory the more minute and circumstantial proof. Thus it
avoids a mysterious, unintelligible phraseology, and makes its progress
in plain language, that is, with a clear and always visible chain of
ideas.

Certainly this cannot always be completely attained, but it must
always be the aim in critical expositions. Such expositions must use
complicated forms of science as sparingly as possible, and never resort
to the construction of scientific aids as of a truth apparatus of its
own, but always be guided by the natural and unbiassed impressions of
the mind.

But this pious endeavour, if we may use the expression, has
unfortunately seldom hitherto presided over critical examinations: the
most of them have rather been emanations of a species of vanity--a wish
to make a display of ideas.

The first evil which we constantly stumble upon is a lame, totally
inadmissible application of certain one-sided systems as of a formal
code of laws. But it is never difficult to show the one-sidedness of
such systems, and this only requires to be done once to throw discredit
for ever on critical judgments which are based on them. We have here
to deal with a definite subject, and as the number of possible systems
after all can be but small, therefore also they are themselves the
lesser evil.

Much greater is the evil which lies in the pompous retinue of technical
terms--scientific expressions and metaphors, which these systems carry
in their train, and which like a rabble-like the baggage of an Army
broken away from its Chief--hang about in all directions. Any critic who
has not adopted a system, either because he has not found one to please
him, or because he has not yet been able to make himself master of one,
will at least occasionally make use of a piece of one, as one would use
a ruler, to show the blunders committed by a General. The most of them
are incapable of reasoning without using as a help here and there some
shreds of scientific military theory. The smallest of these fragments,

consisting in mere scientific words and metaphors, are often nothing
more than ornamental flourishes of critical narration. Now it is in the
nature of things that all technical and scientific expressions which
belong to a system lose their propriety, if they ever had any, as
soon as they are distorted, and used as general axioms, or as small
crystalline talismans, which have more power of demonstration than
simple speech.

Thus it has come to pass that our theoretical and critical books,
instead of being straightforward, intelligible dissertations, in which
the author always knows at least what he says and the reader what he
reads, are brimful of these technical terms, which form dark points of
interference where author and reader part company. But frequently they
are something worse, being nothing but hollow shells without any kernel.
The author himself has no clear perception of what he means, contents
himself with vague ideas, which if expressed in plain language would be
unsatisfactory even to himself.

A third fault in criticism is the MISUSE of HISTORICAL EXAMPLES, and a
display of great reading or learning. What the history of the Art of
War is we have already said, and we shall further explain our views on
examples and on military history in general in special chapters. One
fact merely touched upon in a very cursory manner may be used to support
the most opposite views, and three or four such facts of the most
heterogeneous description, brought together out of the most distant
lands and remote times and heaped up, generally distract and bewilder
the judgment and understanding without demonstrating anything; for when
exposed to the light they turn out to be only trumpery rubbish, made use
of to show off the author's learning.

But what can be gained for practical life by such obscure, partly false,
confused arbitrary conceptions? So little is gained that theory on
account of them has always been a true antithesis of practice, and
frequently a subject of ridicule to those whose soldierly qualities in
the field are above question.

But it is impossible that this could have been the case, if theory
in simple language, and by natural treatment of those things which
constitute the Art of making War, had merely sought to establish just so
much as admits of being established; if, avoiding all false pretensions
and irrelevant display of scientific forms and historical parallels, it
had kept close to the subject, and gone hand in hand with those who must
conduct affairs in the field by their own natural genius.

CHAPTER VI. ON EXAMPLES

EXAMPLES from history make everything clear, and furnish the best
description of proof in the empirical sciences. This applies with more
force to the Art of War than to any other. General Scharnhorst, whose
handbook is the best ever written on actual War, pronounces historical
examples to be of the first importance, and makes an admirable use of
them himself. Had he survived the War in which he fell,(*) the fourth
part of his revised treatise on artillery would have given a still
greater proof of the observing and enlightened spirit in which he sifted
matters of experience.

But such use of historical examples is rarely made by theoretical
writers; the way in which they more commonly make use of them is rather
calculated to leave the mind unsatisfied, as well as to offend the
understanding. We therefore think it important to bring specially into
view the use and abuse of historical examples.

    (*) General Scharnhorst died in 1813, of a wound received in
    the battle of Bautzen or Grosz Gorchen--EDITOR.

Unquestionably the branches of knowledge which lie at the foundation of

the Art of War come under the denomination of empirical sciences; for
although they are derived in a great measure from the nature of things,
still we can only learn this very nature itself for the most part from
experience; and besides that, the practical application is modified by
so many circumstances that the effects can never be completely learnt
from the mere nature of the means.

The effects of gunpowder, that great agent in our military activity,
were only learnt by experience, and up to this hour experiments are
continually in progress in order to investigate them more fully. That an
iron ball to which powder has given a velocity of 1000 feet in a
second, smashes every living thing which it touches in its course is
intelligible in itself; experience is not required to tell us that; but
in producing this effect how many hundred circumstances are concerned,
some of which can only be learnt by experience! And the physical is not
the only effect which we have to study, it is the moral which we are in
search of, and that can only be ascertained by experience; and there is
no other way of learning and appreciating it but by experience. In the
middle ages, when firearms were first invented, their effect, owing to
their rude make, was materially but trifling compared to what it now is,
but their effect morally was much greater. One must have witnessed the
firmness of one of those masses taught and led by Buonaparte, under the
heaviest and most unintermittent cannonade, in order to understand what
troops, hardened by long practice in the field of danger, can do,
when by a career of victory they have reached the noble principle of
demanding from themselves their utmost efforts. In pure conception no
one would believe it. On the other hand, it is well known that there are
troops in the service of European Powers at the present moment who would
easily be dispersed by a few cannon shots.

But no empirical science, consequently also no theory of the Art of War,
can always corroborate its truths by historical proof; it would also be,
in some measure, difficult to support experience by single facts. If
any means is once found efficacious in War, it is repeated; one nation
copies another, the thing becomes the fashion, and in this manner it
comes into use, supported by experience, and takes its place in theory,
which contents itself with appealing to experience in general in order
to show its origin, but not as a verification of its truth.

But it is quite otherwise if experience is to be used in order to
overthrow some means in use, to confirm what is doubtful, or introduce
something new; then particular examples from history must be quoted as
proofs.

Now, if we consider closely the use of historical proofs, four points of
view readily present themselves for the purpose.

First, they may be used merely as an EXPLANATION of an idea. In every
abstract consideration it is very easy to be misunderstood, or not to
be intelligible at all: when an author is afraid of this, an
exemplification from history serves to throw the light which is wanted
on his idea, and to ensure his being intelligible to his reader.

Secondly, it may serve as an APPLICATION of an idea, because by means of
an example there is an opportunity of showing the action of those minor
circumstances which cannot all be comprehended and explained in any
general expression of an idea; for in that consists, indeed, the
difference between theory and experience. Both these cases belong to
examples properly speaking, the two following belong to historical
proofs.

Thirdly, a historical fact may be referred to particularly, in order to
support what one has advanced. This is in all cases sufficient, if we
have ONLY to prove the POSSIBILITY of a fact or effect.

Lastly, in the fourth place, from the circumstantial detail of a
historical event, and by collecting together several of them, we may
deduce some theory, which therefore has its true PROOF in this testimony

itself.

For the first of these purposes all that is generally required is a
cursory notice of the case, as it is only used partially. Historical
correctness is a secondary consideration; a case invented might also
serve the purpose as well, only historical ones are always to be
preferred, because they bring the idea which they illustrate nearer to
practical life.

The second use supposes a more circumstantial relation of events, but
historical authenticity is again of secondary importance, and in respect
to this point the same is to be said as in the first case.

For the third purpose the mere quotation of an undoubted fact is
generally sufficient. If it is asserted that fortified positions may
fulfil their object under certain conditions, it is only necessary to
mention the position of Bunzelwitz(*) in support of the assertion.

    (*) Frederick the Great's celebrated entrenched camp in
    1761.

But if, through the narrative of a case in history, an abstract truth
is to be demonstrated, then everything in the case bearing on the
demonstration must be analysed in the most searching and complete
manner; it must, to a certain extent, develop itself carefully before
the eyes of the reader. The less effectually this is done the weaker
will be the proof, and the more necessary it will be to supply the
demonstrative proof which is wanting in the single case by a number of
cases, because we have a right to suppose that the more minute details
which we are unable to give neutralise each other in their effects in a
certain number of cases.

If we want to show by example derived from experience that cavalry
are better placed behind than in a line with infantry; that it is very
hazardous without a decided preponderance of numbers to attempt an
enveloping movement, with widely separated columns, either on a field
of battle or in the theatre of war--that is, either tactically or
strategically--then in the first of these cases it would not be
sufficient to specify some lost battles in which the cavalry was on the
flanks and some gained in which the cavalry was in rear of the infantry;
and in the tatter of these cases it is not sufficient to refer to the
battles of Rivoli and Wagram, to the attack of the Austrians on the
theatre of war in Italy, in 1796, or of the French upon the German
theatre of war in the same year. The way in which these orders of battle
or plans of attack essentially contributed to disastrous issues in those
particular cases must be shown by closely tracing out circumstances and
occurrences. Then it will appear how far such forms or measures are to
be condemned, a point which it is very necessary to show, for a total
condemnation would be inconsistent with truth.

It has been already said that when a circumstantial detail of facts is
impossible, the demonstrative power which is deficient may to a certain
extent be supplied by the number of cases quoted; but this is a very
dangerous method of getting out of the difficulty, and one which has
been much abused. Instead of one well-explained example, three or four
are just touched upon, and thus a show is made of strong evidence. But
there are matters where a whole dozen of cases brought forward would
prove nothing, if, for instance, they are facts of frequent occurrence,
and therefore a dozen other cases with an opposite result might just as
easily be brought forward. If any one will instance a dozen lost battles
in which the side beaten attacked in separate converging columns, we
can instance a dozen that have been gained in which the same order was
adopted. It is evident that in this way no result is to be obtained.

Upon carefully considering these different points, it will be seen how
easily examples may be misapplied.

An occurrence which, instead of being carefully analysed in all its

parts, is superficially noticed, is like an object seen at a great
distance, presenting the same appearance on each side, and in which the
details of its parts cannot be distinguished. Such examples have, in
reality, served to support the most contradictory opinions. To some
Daun's campaigns are models of prudence and skill. To others, they are
nothing but examples of timidity and want of resolution. Buonaparte's
passage across the Noric Alps in 1797 may be made to appear the noblest
resolution, but also as an act of sheer temerity. His strategic defeat
in 1812 may be represented as the consequence either of an excess, or of
a deficiency, of energy. All these opinions have been broached, and
it is easy to see that they might very well arise, because each person
takes a different view of the connection of events. At the same time
these antagonistic opinions cannot be reconciled with each other, and
therefore one of the two must be wrong.

Much as we are obliged to the worthy Feuquieres for the numerous
examples introduced in his memoirs--partly because a number of
historical incidents have thus been preserved which might otherwise
have been lost, and partly because he was one of the first to bring
theoretical, that is, abstract, ideas into connection with the practical
in war, in so far that the cases brought forward may be regarded as
intended to exemplify and confirm what is theoretically asserted--yet,
in the opinion of an impartial reader, he will hardly be allowed to have
attained the object he proposed to himself, that of proving theoretical
principles by historical examples. For although he sometimes relates
occurrences with great minuteness, still he falls short very often of
showing that the deductions drawn necessarily proceed from the inner
relations of these events.

Another evil which comes from the superficial notice of historical
events, is that some readers are either wholly ignorant of the events,
or cannot call them to remembrance sufficiently to be able to grasp
the author's meaning, so that there is no alternative between either
accepting blindly what is said, or remaining unconvinced.

It is extremely difficult to put together or unfold historical events
before the eyes of a reader in such a way as is necessary, in order
to be able to use them as proofs; for the writer very often wants the
means, and can neither afford the time nor the requisite space; but
we maintain that, when the object is to establish a new or doubtful
opinion, one single example, thoroughly analysed, is far more
instructive than ten which are superficially treated. The great mischief
of these superficial representations is not that the writer puts his
story forward as a proof when it has only a false title, but that he
has not made himself properly acquainted with the subject, and that from
this sort of slovenly, shallow treatment of history, a hundred false
views and attempts at the construction of theories arise, which would
never have made their appearance if the writer had looked upon it as his
duty to deduce from the strict connection of events everything new which
he brought to market, and sought to prove from history.

When we are convinced of these difficulties in the use of historical
examples, and at the same time of the necessity (of making use of such
examples), then we shall also come to the conclusion that the latest
military history is naturally the best field from which to draw them,
inasmuch as it alone is sufficiently authentic and detailed.

In ancient times, circumstances connected with War, as well as the
method of carrying it on, were different; therefore its events are
of less use to us either theoretically or practically; in addition to
which, military history, like every other, naturally loses in the course
of time a number of small traits and lineaments originally to be seen,
loses in colour and life, like a worn-out or darkened picture; so that
perhaps at last only the large masses and leading features remain, which
thus acquire undue proportions.

If we look at the present state of warfare, we should say that the Wars
since that of the Austrian succession are almost the only ones which,

at least as far as armament, have still a considerable similarity to
the present, and which, notwithstanding the many important changes which
have taken place both great and small, are still capable of affording
much instruction. It is quite otherwise with the War of the Spanish
succession, as the use of fire-arms had not then so far advanced towards
perfection, and cavalry still continued the most important arm. The
farther we go back, the less useful becomes military history, as it gets
so much the more meagre and barren of detail. The most useless of all is
that of the old world.

But this uselessness is not altogether absolute, it relates only to
those subjects which depend on a knowledge of minute details, or on
those things in which the method of conducting war has changed. Although
we know very little about the tactics in the battles between the Swiss
and the Austrians, the Burgundians and French, still we find in them
unmistakable evidence that they were the first in which the superiority
of a good infantry over the best cavalry was, displayed. A general
glance at the time of the Condottieri teaches us how the whole method
of conducting war is dependent on the instrument used; for at no period
have the forces used in War had so much the characteristics of a special
instrument, and been a class so totally distinct from the rest of the
national community. The memorable way in which the Romans in the second
Punic War attacked the Carthaginan possessions in Spain and Africa,
while Hannibal still maintained himself in Italy, is a most instructive
subject to study, as the general relations of the States and Armies
concerned in this indirect act of defence are sufficiently well known.

But the more things descend into particulars and deviate in character
from the most general relations, the less we can look for examples and
lessons of experience from very remote periods, for we have neither the
means of judging properly of corresponding events, nor can we apply them
to our completely different method of War.

Unfortunately, however, it has always been the fashion with historical
writers to talk about ancient times. We shall not say how far vanity
and charlatanism may have had a share in this, but in general we fail
to discover any honest intention and earnest endeavour to instruct
and convince, and we can therefore only look upon such quotations and
references as embellishments to fill up gaps and hide defects.

It would be an immense service to teach the Art of War entirely by
historical examples, as Feuquieres proposed to do; but it would be full
work for the whole life of a man, if we reflect that he who undertakes
it must first qualify himself for the task by a long personal experience
in actual War.

Whoever, stirred by ambition, undertakes such a task, let him prepare
himself for his pious undertaking as for a long pilgrimage; let him give
up his time, spare no sacrifice, fear no temporal rank or power, and
rise above all feelings of personal vanity, of false shame, in order,
according to the French code, to speak THE TRUTH, THE WHOLE TRUTH, AND
NOTHING BUT THE TRUTH.

BOOK III. OF STRATEGY IN GENERAL

CHAPTER I. STRATEGY

IN the second chapter of the second book, Strategy has been defined as
"the employment of the battle as the means towards the attainment of the
object of the war." Properly speaking it has to do with nothing but the
battle, but its theory must include in this consideration the instrument
of this real activity--the armed force--in itself and in its principal
relations, for the battle is fought by it, and shows its effects upon
it in turn. It must be well acquainted with the battle itself as far as

relates to its possible results, and those mental and moral powers which
are the most important in the use of the same.

Strategy is the employment of the battle to gain the end of the War; it
must therefore give an aim to the whole military action, which must be
in accordance with the object of the War; in other words, Strategy forms
the plan of the War, and to this end it links together the series of
acts which are to lead to the final decision, that, is to say, it makes
the plans for the separate campaigns and regulates the combats to be
fought in each. As these are all things which to a great extent can only
be determined on conjectures some of which turn out incorrect, while a
number of other arrangements pertaining to details cannot be made at
all beforehand, it follows, as a matter of course, that Strategy must go
with the Army to the field in order to arrange particulars on the spot,
and to make the modifications in the general plan, which incessantly
become necessary in War. Strategy can therefore never take its hand from
the work for a moment.

That this, however, has not always been the view taken is evident from
the former custom of keeping Strategy in the cabinet and not with the
Army, a thing only allowable if the cabinet is so near to the Army that
it can be taken for the chief head-quarters of the Army.

Theory will therefore attend on Strategy in the determination of its
plans, or, as we may more properly say, it will throw a light on things
in themselves, and on their relations to each other, and bring out
prominently the little that there is of principle or rule.

If we recall to mind from the first chapter how many things of
the highest importance War touches upon, we may conceive that a
consideration of all requires a rare grasp of mind.

A Prince or General who knows exactly how to organise his War according
to his object and means, who does neither too little nor too much, gives
by that the greatest proof of his genius. But the effects of this talent
are exhibited not so much by the invention of new modes of action, which
might strike the eye immediately, as in the successful final result of
the whole. It is the exact fulfilment of silent suppositions, it is the
noiseless harmony of the whole action which we should admire, and which
only makes itself known in the total result. Inquirer who, tracing back
from the final result, does not perceive the signs of that harmony is
one who is apt to seek for genius where it is not, and where it cannot
be found.

The means and forms which Strategy uses are in fact so extremely
simple, so well known by their constant repetition, that it only appears
ridiculous to sound common sense when it hears critics so frequently
speaking of them with high-flown emphasis. Turning a flank, which has
been done a thousand times, is regarded here as a proof of the most
brilliant genius, there as a proof of the most profound penetration,
indeed even of the most comprehensive knowledge. Can there be in the
book-world more absurd productions?(*)

> (*) This paragraph refers to the works of Lloyd, Buelow,
> indeed to all the eighteenth-century writers, from whose
> influence we in England are not even yet free.--ED.

It is still more ridiculous if, in addition to this, we reflect that the
same critic, in accordance with prevalent opinion, excludes all moral
forces from theory, and will not allow it to be concerned with
anything but the material forces, so that all must be confined to a few
mathematical relations of equilibrium and preponderance, of time and
space, and a few lines and angles. If it were nothing more than this,
then out of such a miserable business there would not be a scientific
problem for even a schoolboy.

But let us admit: there is no question here about scientific formulas
and problems; the relations of material things are all very simple; the

right comprehension of the moral forces which come into play is more difficult. Still, even in respect to them, it is only in the highest branches of Strategy that moral complications and a great diversity of quantities and relations are to be looked for, only at that point where Strategy borders on political science, or rather where the two become one, and there, as we have before observed, they have more influence on the "how much" and "how little" is to be done than on the form of execution. Where the latter is the principal question, as in the single acts both great and small in War, the moral quantities are already reduced to a very small number.

Thus, then, in Strategy everything is very simple, but not on that account very easy. Once it is determined from the relations of the State what should and may be done by War, then the way to it is easy to find; but to follow that way straightforward, to carry out the plan without being obliged to deviate from it a thousand times by a thousand varying influences, requires, besides great strength of character, great clearness and steadiness of mind, and out of a thousand men who are remarkable, some for mind, others for penetration, others again for boldness or strength of will, perhaps not one will combine in himself all those qualities which are required to raise a man above mediocrity in the career of a general.

It may sound strange, but for all who know War in this respect it is a fact beyond doubt, that much more strength of will is required to make an important decision in Strategy than in tactics. In the latter we are hurried on with the moment; a Commander feels himself borne along in a strong current, against which he durst not contend without the most destructive consequences, he suppresses the rising fears, and boldly ventures further. In Strategy, where all goes on at a slower rate, there is more room allowed for our own apprehensions and those of others, for objections and remonstrances, consequently also for unseasonable regrets; and as we do not see things in Strategy as we do at least half of them in tactics, with the living eye, but everything must be conjectured and assumed, the convictions produced are less powerful. The consequence is that most Generals, when they should act, remain stuck fast in bewildering doubts.

Now let us cast a glance at history--upon Frederick the Great's campaign of 1760, celebrated for its fine marches and manoeuvres: a perfect masterpiece of Strategic skill as critics tell us. Is there really anything to drive us out of our wits with admiration in the King's first trying to turn Daun's right flank, then his left, then again his right, &c.? Are we to see profound wisdom in this? No, that we cannot, if we are to decide naturally and without affectation. What we rather admire above all is the sagacity of the King in this respect, that while pursuing a great object with very limited means, he undertook nothing beyond his powers, and JUST ENOUGH to gain his object. This sagacity of the General is visible not only in this campaign, but throughout all the three Wars of the Great King!

To bring Silesia into the safe harbour of a well-guaranteed peace was his object.

At the head of a small State, which was like other States in most things, and only ahead of them in some branches of administration; he could not be an Alexander, and, as Charles XII, he would only, like him, have broken his head. We find, therefore, in the whole of his conduct of War, a controlled power, always well balanced, and never wanting in energy, which in the most critical moments rises to astonishing deeds, and the next moment oscillates quietly on again in subordination to the play of the most subtle political influences. Neither vanity, thirst for glory, nor vengeance could make him deviate from his course, and this course alone it is which brought him to a fortunate termination of the contest.

These few words do but scant justice to this phase of the genius of the great General; the eyes must be fixed carefully on the extraordinary issue of the struggle, and the causes which brought about that issue must be traced out, in order thoroughly to understand that nothing but the King's penetrating eye brought him safely out of all his dangers.

This is one feature in this great Commander which we admire in the campaign of 1760--and in all others, but in this especially--because in none did he keep the balance even against such a superior hostile force, with such a small sacrifice.

Another feature relates to the difficulty of execution. Marches to turn a flank, right or left, are easily combined; the idea of keeping a small force always well concentrated to be able to meet the enemy on equal terms at any point, to multiply a force by rapid movement, is as easily conceived as expressed; the mere contrivance in these points, therefore, cannot excite our admiration, and with respect to such simple things, there is nothing further than to admit that they are simple.

But let a General try to do these things like Frederick the Great. Long afterwards authors, who were eyewitnesses, have spoken of the danger, indeed of the imprudence, of the King's camps, and doubtless, at the time he pitched them, the danger appeared three times as great as afterwards.

It was the same with his marches, under the eyes, nay, often under the cannon of the enemy's Army; these camps were taken up, these marches made, not from want of prudence, but because in Daun's system, in his mode of drawing up his Army, in the responsibility which pressed upon him, and in his character, Frederick found that security which justified his camps and marches. But it required the King's boldness, determination, and strength of will to see things in this light, and not to be led astray and intimidated by the danger of which thirty years after people still wrote and spoke. Few Generals in this situation would have believed these simple strategic means to be practicable.

Again, another difficulty in execution lay in this, that the King's Army in this campaign was constantly in motion. Twice it marched by wretched cross-roads, from the Elbe into Silesia, in rear of Daun and pursued by Lascy (beginning of July, beginning of August). It required to be always ready for battle, and its marches had to be organised with a degree of skill which necessarily called forth a proportionate amount of exertion. Although attended and delayed by thousands of waggons, still its subsistence was extremely difficult. In Silesia, for eight days before the battle of Leignitz, it had constantly to march, defiling alternately right and left in front of the enemy:--this costs great fatigue, and entails great privations.

Is it to be supposed that all this could have been done without producing great friction in the machine? Can the mind of a Commander elaborate such movements with the same ease as the hand of a land surveyor uses the astrolabe? Does not the sight of the sufferings of their hungry, thirsty comrades pierce the hearts of the Commander and his Generals a thousand times? Must not the murmurs and doubts which these cause reach his ear? Has an ordinary man the courage to demand such sacrifices, and would not such efforts most certainly demoralise the Army, break up the bands of discipline, and, in short, undermine its military virtue, if firm reliance on the greatness and infallibility of the Commander did not compensate for all? Here, therefore, it is that we should pay respect; it is these miracles of execution which we should admire. But it is impossible to realise all this in its full force without a foretaste of it by experience. He who only knows War from books or the drill-ground cannot realise the whole effect of this counterpoise in action; WE BEG HIM, THEREFORE, TO ACCEPT FROM US ON FAITH AND TRUST ALL THAT HE IS UNABLE TO SUPPLY FROM ANY PERSONAL EXPERIENCES OF HIS OWN.

This illustration is intended to give more clearness to the course of our ideas, and in closing this chapter we will only briefly observe that in our exposition of Strategy we shall describe those separate subjects

which appear to us the most important, whether of a moral or material
nature; then proceed from the simple to the complex, and conclude with
the inner connection of the whole act of War, in other words, with the
plan for a War or campaign.

OBSERVATION.

In an earlier manuscript of the second book are the following passages
endorsed by the author himself to be used for the first Chapter of the
second Book: the projected revision of that chapter not having been
made, the passages referred to are introduced here in full.

By the mere assemblage of armed forces at a particular point, a
battle there becomes possible, but does not always take place. Is that
possibility now to be regarded as a reality and therefore an effective
thing? Certainly, it is so by its results, and these effects, whatever
they may be, can never fail.

1. POSSIBLE COMBATS ARE ON ACCOUNT OF THEIR RESULTS TO BE LOOKED UPON AS
REAL ONES.

If a detachment is sent away to cut off the retreat of a flying enemy,
and the enemy surrenders in consequence without further resistance,
still it is through the combat which is offered to him by this
detachment sent after him that he is brought to his decision.

If a part of our Army occupies an enemy's province which was undefended,
and thus deprives the enemy of very considerable means of keeping up
the strength of his Army, it is entirely through the battle which our
detached body gives the enemy to expect, in case he seeks to recover the
lost province, that we remain in possession of the same.

In both cases, therefore, the mere possibility of a battle has produced
results, and is therefore to be classed amongst actual events. Suppose
that in these cases the enemy has opposed our troops with others
superior in force, and thus forced ours to give up their object without
a combat, then certainly our plan has failed, but the battle which we
offered at (either of) those points has not on that account been without
effect, for it attracted the enemy's forces to that point. And in case
our whole undertaking has done us harm, it cannot be said that these
positions, these possible battles, have been attended with no results;
their effects, then, are similar to those of a lost battle.

In this manner we see that the destruction of the enemy's military
forces, the overthrow of the enemy's power, is only to be done through
the effect of a battle, whether it be that it actually takes place, or
that it is merely offered, and not accepted.

2. TWOFOLD OBJECT OF THE COMBAT.

But these effects are of two kinds, direct and indirect they are of the
latter, if other things intrude themselves and become the object of the
combat--things which cannot be regarded as the destruction of enemy's
force, but only leading up to it, certainly by a circuitous road, but
with so much the greater effect. The possession of provinces, towns,
fortresses, roads, bridges, magazines, &c., may be the IMMEDIATE object
of a battle, but never the ultimate one. Things of this description
can never be, looked upon otherwise than as means of gaining greater
superiority, so as at last to offer battle to the enemy in such a way
that it will be impossible for him to accept it. Therefore all these
things must only be regarded as intermediate links, steps, as it were,
leading up to the effectual principle, but never as that principle
itself.

3. EXAMPLE.

In 1814, by the capture of Buonaparte's capital the object of the War
was attained. The political divisions which had their roots in Paris
came into active operation, and an enormous split left the power of the
Emperor to collapse of itself. Nevertheless the point of view from which
we must look at all this is, that through these causes the forces and
defensive means of Buonaparte were suddenly very much diminished,
the superiority of the Allies, therefore, just in the same measure
increased, and any further resistance then became IMPOSSIBLE. It was
this impossibility which produced the peace with France. If we suppose
the forces of the Allies at that moment diminished to a like extent
through external causes;--if the superiority vanishes, then at the same
time vanishes also all the effect and importance of the taking of Paris.

We have gone through this chain of argument in order to show that this
is the natural and only true view of the thing from which it derives
its importance. It leads always back to the question, what at any given
moment of the War or campaign will be the probable result of the great
or small combats which the two sides might offer to each other? In the
consideration of a plan for a campaign, this question only is decisive
as to the measures which are to be taken all through from the very
commencement.

4. WHEN THIS VIEW IS NOT TAKEN, THEN A FALSE VALUE IS GIVEN TO OTHER
THINGS.

If we do not accustom ourselves to look upon War, and the single
campaigns in a War, as a chain which is all composed of battles strung
together, one of which always brings on another; if we adopt the idea
that the taking of a certain geographical point, the occupation of an
undefended province, is in itself anything; then we are very likely to
regard it as an acquisition which we may retain; and if we look at
it so, and not as a term in the whole series of events, we do not ask
ourselves whether this possession may not lead to greater disadvantages
hereafter. How often we find this mistake recurring in military history.

We might say that, just as in commerce the merchant cannot set apart and
place in security gains from one single transaction by itself, so in
War a single advantage cannot be separated from the result of the whole.
Just as the former must always operate with the whole bulk of his means,
just so in War, only the sum total will decide on the advantage or
disadvantage of each item.

If the mind's eye is always directed upon the series of combats, so far
as they can be seen beforehand, then it is always looking in the right
direction, and thereby the motion of the force acquires that rapidity,
that is to say, willing and doing acquire that energy which is suitable
to the matter, and which is not to be thwarted or turned aside by
extraneous influences.(*)

    (*) The whole of this chapter is directed against the
    theories of the Austrian Staff in 1814. It may be taken as
    the foundation of the modern teaching of the Prussian
    General Staff. See especially von Kammer.--ED.

CHAPTER II. ELEMENTS OF STRATEGY

THE causes which condition the use of the combat in Strategy may be
easily divided into elements of different kinds, such as the moral,
physical, mathematical, geographical and statistical elements.

The first class includes all that can be called forth by moral qualities
and effects; to the second belong the whole mass of the military force,
its organisation, the proportion of the three arms, &c. &c.; to the

third, the angle of the lines of operation, the concentric and eccentric
movements in as far as their geometrical nature has any value in
the calculation; to the fourth, the influences of country, such as
commanding points, hills, rivers, woods, roads, &c. &c.; lastly, to the
fifth, all the means of supply. The separation of these things once for
all in the mind does good in giving clearness and helping us to estimate
at once, at a higher or lower value, the different classes as we pass
onwards. For, in considering them separately, many lose of themselves
their borrowed importance; one feels, for instance, quite plainly that
the value of a base of operations, even if we look at nothing in it but
its relative position to the line of operations, depends much less in
that simple form on the geometrical element of the angle which they
form with one another, than on the nature of the roads and the country
through which they pass.

But to treat upon Strategy according to these elements would be the
most unfortunate idea that could be conceived, for these elements are
generally manifold, and intimately connected with each other in every
single operation of War. We should lose ourselves in the most soulless
analysis, and as if in a horrid dream, we should be for ever trying in
vain to build up an arch to connect this base of abstractions with facts
belonging to the real world. Heaven preserve every theorist from such an
undertaking! We shall keep to the world of things in their totality, and
not pursue our analysis further than is necessary from time to time to
give distinctness to the idea which we wish to impart, and which
has come to us, not by a speculative investigation, but through the
impression made by the realities of War in their entirety.

CHAPTER III. MORAL FORCES

WE must return again to this subject, which is touched upon in the third
chapter of the second book, because the moral forces are amongst the
most important subjects in War. They form the spirit which permeates the
whole being of War. These forces fasten themselves soonest and with the
greatest affinity on to the Will which puts in motion and guides the
whole mass of powers, uniting with it as it were in one stream, because
this is a moral force itself. Unfortunately they will escape from all
book-analysis, for they will neither be brought into numbers nor into
classes, and require to be both seen and felt.

The spirit and other moral qualities which animate an Army, a General,
or Governments, public opinion in provinces in which a War is raging,
the moral effect of a victory or of a defeat, are things which in
themselves vary very much in their nature, and which also, according
as they stand with regard to our object and our relations, may have an
influence in different ways.

Although little or nothing can be said about these things in books,
still they belong to the theory of the Art of War, as much as everything
else which constitutes War. For I must here once more repeat that it is
a miserable philosophy if, according to the old plan, we establish rules
and principles wholly regardless of all moral forces, and then, as soon
as these forces make their appearance, we begin to count exceptions
which we thereby establish as it were theoretically, that is, make into
rules; or if we resort to an appeal to genius, which is above all rules,
thus giving out by implication, not only that rules were only made for
fools, but also that they themselves are no better than folly.

Even if the theory of the Art of War does no more in reality than recall
these things to remembrance, showing the necessity of allowing to
the moral forces their full value, and of always taking them into
consideration, by so doing it extends its borders over the region of
immaterial forces, and by establishing that point of view, condemns
beforehand every one who would endeavour to justify himself before its
judgment seat by the mere physical relations of forces.

Further out of regard to all other so-called rules, theory cannot

banish the moral forces beyond its frontier, because the effects of the
physical forces and the moral are completely fused, and are not to
be decomposed like a metal alloy by a chemical process. In every rule
relating to the physical forces, theory must present to the mind at the
same time the share which the moral powers will have in it, if it
would not be led to categorical propositions, at one time too timid
and contracted, at another too dogmatical and wide. Even the most
matter-of-fact theories have, without knowing it, strayed over into this
moral kingdom; for, as an example, the effects of a victory cannot
in any way be explained without taking into consideration the moral
impressions. And therefore the most of the subjects which we shall go
through in this book are composed half of physical, half of moral causes
and effects, and we might say the physical are almost no more than
the wooden handle, whilst the moral are the noble metal, the real
bright-polished weapon.

The value of the moral powers, and their frequently incredible
influence, are best exemplified by history, and this is the most
generous and the purest nourishment which the mind of the General can
extract from it.--At the same time it is to be observed, that it is
less demonstrations, critical examinations, and learned treatises, than
sentiments, general impressions, and single flashing sparks of truth,
which yield the seeds of knowledge that are to fertilise the mind.

We might go through the most important moral phenomena in War, and with
all the care of a diligent professor try what we could impart about
each, either good or bad. But as in such a method one slides too much
into the commonplace and trite, whilst real mind quickly makes its
escape in analysis, the end is that one gets imperceptibly to the
relation of things which everybody knows. We prefer, therefore, to
remain here more than usually incomplete and rhapsodical, content to
have drawn attention to the importance of the subject in a general way,
and to have pointed out the spirit in which the views given in this book
have been conceived.

CHAPTER IV. THE CHIEF MORAL POWERS

THESE are The Talents of the Commander; The Military Virtue of the Army;
Its National feeling. Which of these is the most important no one can
tell in a general way, for it is very difficult to say anything in
general of their strength, and still more difficult to compare the
strength of one with that of another. The best plan is not to undervalue
any of them, a fault which human judgment is prone to, sometimes on one
side, sometimes on another, in its whimsical oscillations. It is better
to satisfy ourselves of the undeniable efficacy of these three things by
sufficient evidence from history.

It is true, however, that in modern times the Armies of European states
have arrived very much at a par as regards discipline and fitness
for service, and that the conduct of War has--as philosophers would
say--naturally developed itself, thereby become a method, common as
it were to all Armies, so that even from Commanders there is nothing
further to be expected in the way of application of special means
of Art, in the limited sense (such as Frederick the Second's oblique
order). Hence it cannot be denied that, as matters now stand, greater
scope is afforded for the influence of National spirit and habituation
of an army to War. A long peace may again alter all this.(*)

          (*) Written shortly after the Great Napoleonic campaigns.

The national spirit of an Army (enthusiasm, fanatical zeal, faith,
opinion) displays itself most in mountain warfare, where every one down
to the common soldier is left to himself. On this account, a mountainous
country is the best campaigning ground for popular levies.

Expertness of an Army through training, and that well-tempered courage which holds the ranks together as if they had been cast in a mould, show their superiority in an open country.

The talent of a General has most room to display itself in a closely intersected, undulating country. In mountains he has too little command over the separate parts, and the direction of all is beyond his powers; in open plains it is simple and does not exceed those powers.

According to these undeniable elective affinities, plans should be regulated.

CHAPTER V. MILITARY VIRTUE OF AN ARMY

THIS is distinguished from mere bravery, and still more from enthusiasm for the business of War. The first is certainly a necessary constituent part of it, but in the same way as bravery, which is a natural gift in some men, may arise in a soldier as a part of an Army from habit and custom, so with him it must also have a different direction from that which it has with others. It must lose that impulse to unbridled activity and exercise of force which is its characteristic in the individual, and submit itself to demands of a higher kind, to obedience, order, rule, and method. Enthusiasm for the profession gives life and greater fire to the military virtue of an Army, but does not necessarily constitute a part of it.

War is a special business, and however general its relations may be, and even if all the male population of a country, capable of bearing arms, exercise this calling, still it always continues to be different and separate from the other pursuits which occupy the life of man.--To be imbued with a sense of the spirit and nature of this business, to make use of, to rouse, to assimilate into the system the powers which should be active in it, to penetrate completely into the nature of the business with the understanding, through exercise to gain confidence and expertness in it, to be completely given up to it, to pass out of the man into the part which it is assigned to us to play in War, that is the military virtue of an Army in the individual.

However much pains may be taken to combine the soldier and the citizen in one and the same individual, whatever may be done to nationalise Wars, and however much we may imagine times have changed since the days of the old Condottieri, never will it be possible to do away with the individuality of the business; and if that cannot be done, then those who belong to it, as long as they belong to it, will always look upon themselves as a kind of guild, in the regulations, laws and customs in which the "Spirit of War" by preference finds its expression. And so it is in fact. Even with the most decided inclination to look at War from the highest point of view, it would be very wrong to look down upon this corporate spirit (e'sprit de corps) which may and should exist more or less in every Army. This corporate spirit forms the bond of union between the natural forces which are active in that which we have called military virtue. The crystals of military virtue have a greater affinity for the spirit of a corporate body than for anything else.

An Army which preserves its usual formations under the heaviest fire, which is never shaken by imaginary fears, and in the face of real danger disputes the ground inch by inch, which, proud in the feeling of its victories, never loses its sense of obedience, its respect for and confidence in its leaders, even under the depressing effects of defeat; an Army with all its physical powers, inured to privations and fatigue by exercise, like the muscles of an athlete; an Army which looks upon all its toils as the means to victory, not as a curse which hovers over its standards, and which is always reminded of its duties and virtues by the short catechism of one idea, namely the HONOUR OF ITS ARMS;--Such an Army is imbued with the true military spirit.

Soldiers may fight bravely like the Vende'ans, and do great things like the Swiss, the Americans, or Spaniards, without displaying this military

virtue. A Commander may also be successful at the head of standing
Armies, like Eugene and Marlborough, without enjoying the benefit of its
assistance; we must not, therefore, say that a successful War without
it cannot be imagined; and we draw especial attention to that point,
in order the more to individualise the conception which is here brought
forward, that the idea may not dissolve into a generalisation and that
it may not be thought that military virtue is in the end everything. It
is not so. Military virtue in an Army is a definite moral power which
may be supposed wanting, and the influence of which may therefore be
estimated--like any instrument the power of which may be calculated.

Having thus characterised it, we proceed to consider what can be
predicated of its influence, and what are the means of gaining its
assistance.

Military virtue is for the parts, what the genius of the Commander is
for the whole. The General can only guide the whole, not each separate
part, and where he cannot guide the part, there military virtue must
be its leader. A General is chosen by the reputation of his superior
talents, the chief leaders of large masses after careful probation; but
this probation diminishes as we descend the scale of rank, and in just
the same measure we may reckon less and less upon individual talents;
but what is wanting in this respect military virtue should supply. The
natural qualities of a warlike people play just this part: BRAVERY,
APTITUDE, POWERS OF ENDURANCE and ENTHUSIASM.

These properties may therefore supply the place of military virtue, and
vice versa, from which the following may be deduced:

1. Military virtue is a quality of standing Armies only, but they
require it the most. In national risings its place is supplied by
natural qualities, which develop themselves there more rapidly.

2. Standing Armies opposed to standing Armies, can more easily dispense
with it, than a standing Army opposed to a national insurrection, for in
that case, the troops are more scattered, and the divisions left more
to themselves. But where an Army can be kept concentrated, the genius of
the General takes a greater place, and supplies what is wanting in the
spirit of the Army. Therefore generally military virtue becomes more
necessary the more the theatre of operations and other circumstances
make the War complicated, and cause the forces to be scattered.

From these truths the only lesson to be derived is this, that if an Army
is deficient in this quality, every endeavour should be made to simplify
the operations of the War as much as possible, or to introduce double
efficiency in the organisation of the Army in some other respect, and
not to expect from the mere name of a standing Army, that which only the
veritable thing itself can give.

The military virtue of an Army is, therefore, one of the most important
moral powers in War, and where it is wanting, we either see its
place supplied by one of the others, such as the great superiority
of generalship or popular enthusiasm, or we find the results not
commensurate with the exertions made.--How much that is great, this
spirit, this sterling worth of an army, this refining of ore into
the polished metal, has already done, we see in the history of the
Macedonians under Alexander, the Roman legions under Cesar, the Spanish
infantry under Alexander Farnese, the Swedes under Gustavus Adolphus
and Charles XII, the Prussians under Frederick the Great, and the French
under Buonaparte. We must purposely shut our eyes against all historical
proof, if we do not admit, that the astonishing successes of these
Generals and their greatness in situations of extreme difficulty, were
only possible with Armies possessing this virtue.

This spirit can only be generated from two sources, and only by these
two conjointly; the first is a succession of campaigns and great
victories; the other is, an activity of the Army carried sometimes to
the highest pitch. Only by these, does the soldier learn to know his

powers. The more a General is in the habit of demanding from his troops, the surer he will be that his demands will be answered. The soldier is as proud of overcoming toil, as he is of surmounting danger. Therefore it is only in the soil of incessant activity and exertion that the germ will thrive, but also only in the sunshine of victory. Once it becomes a STRONG TREE, it will stand against the fiercest storms of misfortune and defeat, and even against the indolent inactivity of peace, at least for a time. It can therefore only be created in War, and under great Generals, but no doubt it may last at least for several generations, even under Generals of moderate capacity, and through considerable periods of peace.

With this generous and noble spirit of union in a line of veteran troops, covered with scars and thoroughly inured to War, we must not compare the self-esteem and vanity of a standing Army,(*) held together merely by the glue of service-regulations and a drill book; a certain plodding earnestness and strict discipline may keep up military virtue for a long time, but can never create it; these things therefore have a certain value, but must not be over-rated. Order, smartness, good will, also a certain degree of pride and high feeling, are qualities of an Army formed in time of peace which are to be prized, but cannot stand alone. The whole retains the whole, and as with glass too quickly cooled, a single crack breaks the whole mass. Above all, the highest spirit in the world changes only too easily at the first check into depression, and one might say into a kind of rhodomontade of alarm, the French sauve que peut.--Such an Army can only achieve something through its leader, never by itself. It must be led with double caution, until by degrees, in victory and hardships, the strength grows into the full armour. Beware then of confusing the SPIRIT of an Army with its temper.

   (*) Clausewitz is, of course, thinking of the long-service
   standing armies of his own youth. Not of the short-service
   standing armies of to-day (EDITOR).

CHAPTER VI. BOLDNESS

THE place and part which boldness takes in the dynamic system of powers, where it stands opposed to Foresight and prudence, has been stated in the chapter on the certainty of the result in order thereby to show, that theory has no right to restrict it by virtue of its legislative power.

But this noble impulse, with which the human soul raises itself above the most formidable dangers, is to be regarded as an active principle peculiarly belonging to War. In fact, in what branch of human activity should boldness have a right of citizenship if not in War?

From the transport-driver and the drummer up to the General, it is the noblest of virtues, the true steel which gives the weapon its edge and brilliancy.

Let us admit in fact it has in War even its own prerogatives. Over and above the result of the calculation of space, time, and quantity, we must allow a certain percentage which boldness derives from the weakness of others, whenever it gains the mastery. It is therefore, virtually, a creative power. This is not difficult to demonstrate philosophically. As often as boldness encounters hesitation, the probability of the result is of necessity in its favour, because the very state of hesitation implies a loss of equilibrium already. It is only when it encounters cautious foresight--which we may say is just as bold, at all events just as strong and powerful as itself--that it is at a disadvantage; such cases, however, rarely occur. Out of the whole multitude of prudent men in the world, the great majority are so from timidity.

Amongst large masses, boldness is a force, the special cultivation of which can never be to the detriment of other forces, because the great

mass is bound to a higher will by the frame-work and joints of the order
of battle and of the service, and therefore is guided by an intelligent
power which is extraneous. Boldness is therefore here only like a spring
held down until its action is required.

The higher the rank the more necessary it is that boldness should
be accompanied by a reflective mind, that it may not be a mere blind
outburst of passion to no purpose; for with increase of rank it
becomes always less a matter of self-sacrifice and more a matter of the
preservation of others, and the good of the whole. Where regulations
of the service, as a kind of second nature, prescribe for the masses,
reflection must be the guide of the General, and in his case individual
boldness in action may easily become a fault. Still, at the same time,
it is a fine failing, and must not be looked at in the same light as any
other. Happy the Army in which an untimely boldness frequently manifests
itself; it is an exuberant growth which shows a rich soil. Even
foolhardiness, that is boldness without an object, is not to be
despised; in point of fact it is the same energy of feeling, only
exercised as a kind of passion without any co-operation of the
intelligent faculties. It is only when it strikes at the root of
obedience, when it treats with contempt the orders of superior
authority, that it must be repressed as a dangerous evil, not on its own
account but on account of the act of disobedience, for there is nothing
in War which is of GREATER IMPORTANCE THAN OBEDIENCE.

The reader will readily agree with us that, supposing an equal degree of
discernment to be forthcoming in a certain number of cases, a thousand
times as many of them will end in disaster through over-anxiety as
through boldness.

One would suppose it natural that the interposition of a reasonable
object should stimulate boldness, and therefore lessen its intrinsic
merit, and yet the reverse is the case in reality.

The intervention of lucid thought or the general supremacy of mind
deprives the emotional forces of a great part of their power. On that
account BOLDNESS BECOMES OF RARER OCCURRENCE THE HIGHER WE ASCEND THE
SCALE OF RANK, for whether the discernment and the understanding do or
do not increase with these ranks still the Commanders, in their several
stations as they rise, are pressed upon more and more severely by
objective things, by relations and claims from without, so that they
become the more perplexed the lower the degree of their individual
intelligence. This so far as regards War is the chief foundation of the
truth of the French proverb:--

"Tel brille au second qui s' e'clipse an premier."

Almost all the Generals who are represented in history as merely having
attained to mediocrity, and as wanting in decision when in supreme
command, are men celebrated in their antecedent career for their
boldness and decision.(*)

        (*) Beaulieu, Benedek, Bazaine, Buller, Melas, Mack. &c. &c.

In those motives to bold action which arise from the pressure of
necessity we must make a distinction. Necessity has its degrees of
intensity. If it lies near at hand, if the person acting is in the
pursuit of his object driven into great dangers in order to escape
others equally great, then we can only admire his resolution,
which still has also its value. If a young man to show his skill in
horsemanship leaps across a deep cleft, then he is bold; if he makes
the same leap pursued by a troop of head-chopping Janissaries he is only
resolute. But the farther off the necessity from the point of action,
the greater the number of relations intervening which the mind has to
traverse; in order to realise them, by so much the less does necessity
take from boldness in action. If Frederick the Great, in the year 1756,
saw that War was inevitable, and that he could only escape destruction

by being beforehand with his enemies, it became necessary for him to
commence the War himself, but at the same time it was certainly very
bold: for few men in his position would have made up their minds to do
so.

Although Strategy is only the province of Generals-in-Chief or
Commanders in the higher positions, still boldness in all the other
branches of an Army is as little a matter of indifference to it as their
other military virtues. With an Army belonging to a bold race, and in
which the spirit of boldness has been always nourished, very different
things may be undertaken than with one in which this virtue, is unknown;
for that reason we have considered it in connection with an Army. But
our subject is specially the boldness of the General, and yet we have
not much to say about it after having described this military virtue in
a general way to the best of our ability.

The higher we rise in a position of command, the more of the mind,
understanding, and penetration predominate in activity, the more
therefore is boldness, which is a property of the feelings, kept in
subjection, and for that reason we find it so rarely in the highest
positions, but then, so much the more should it be admired. Boldness,
directed by an overruling intelligence, is the stamp of the hero: this
boldness does not consist in venturing directly against the nature of
things, in a downright contempt of the laws of probability, but, if
a choice is once made, in the rigorous adherence to that higher
calculation which genius, the tact of judgment, has gone over with the
speed of lightning. The more boldness lends wings to the mind and the
discernment, so much the farther they will reach in their flight, so
much the more comprehensive will be the view, the more exact the result,
but certainly always only in the sense that with greater objects greater
dangers are connected. The ordinary man, not to speak of the weak
and irresolute, arrives at an exact result so far as such is possible
without ocular demonstration, at most after diligent reflection in his
chamber, at a distance from danger and responsibility. Let danger and
responsibility draw close round him in every direction, then he loses
the power of comprehensive vision, and if he retains this in any measure
by the influence of others, still he will lose his power of DECISION,
because in that point no one can help him.

We think then that it is impossible to imagine a distinguished General
without boldness, that is to say, that no man can become one who is not
born with this power of the soul, and we therefore look upon it as
the first requisite for such a career. How much of this inborn power,
developed and moderated through education and the circumstances of
life, is left when the man has attained a high position, is the second
question. The greater this power still is, the stronger will genius
be on the wing, the higher will be its flight. The risks become always
greater, but the purpose grows with them. Whether its lines proceed out
of and get their direction from a distant necessity, or whether they
converge to the keystone of a building which ambition has planned,
whether Frederick or Alexander acts, is much the same as regards the
critical view. If the one excites the imagination more because it is
bolder, the other pleases the understanding most, because it has in it
more absolute necessity.

We have still to advert to one very important circumstance.

The spirit of boldness can exist in an Army, either because it is in the
people, or because it has been generated in a successful War conducted
by able Generals. In the latter case it must of course be dispensed with
at the commencement.

Now in our days there is hardly any other means of educating the spirit
of a people in this respect, except by War, and that too under bold
Generals. By it alone can that effeminacy of feeling be counteracted,
that propensity to seek for the enjoyment of comfort, which cause
degeneracy in a people rising in prosperity and immersed in an extremely
busy commerce.

A Nation can hope to have a strong position in the political world only
if its character and practice in actual War mutually support each other
in constant reciprocal action.

CHAPTER VII. PERSEVERANCE

THE reader expects to hear of angles and lines, and finds, instead of
these citizens of the scientific world, only people out of common life,
such as he meets with every day in the street. And yet the author cannot
make up his mind to become a hair's breadth more mathematical than the
subject seems to him to require, and he is not alarmed at the surprise
which the reader may show.

In War more than anywhere else in the world things happen differently to
what we had expected, and look differently when near, to what they
did at a distance. With what serenity the architect can watch his work
gradually rising and growing into his plan. The doctor although much
more at the mercy of mysterious agencies and chances than the architect,
still knows enough of the forms and effects of his means. In War, on
the other hand, the Commander of an immense whole finds himself in a
constant whirlpool of false and true information, of mistakes
committed through fear, through negligence, through precipitation,
of contraventions of his authority, either from mistaken or correct
motives, from ill will, true or false sense of duty, indolence or
exhaustion, of accidents which no mortal could have foreseen. In short,
he is the victim of a hundred thousand impressions, of which the most
have an intimidating, the fewest an encouraging tendency. By long
experience in War, the tact is acquired of readily appreciating the
value of these incidents; high courage and stability of character stand
proof against them, as the rock resists the beating of the waves. He who
would yield to these impressions would never carry out an undertaking,
and on that account PERSEVERANCE in the proposed object, as long as
there is no decided reason against it, is a most necessary counterpoise.
Further, there is hardly any celebrated enterprise in War which was not
achieved by endless exertion, pains, and privations; and as here the
weakness of the physical and moral man is ever disposed to yield, only
an immense force of will, which manifests itself in perseverance admired
by present and future generations, can conduct to our goal.

CHAPTER VIII. SUPERIORITY OF NUMBERS

THIS is in tactics, as well as in Strategy, the most general principle
of victory, and shall be examined by us first in its generality, for
which we may be permitted the following exposition:

Strategy fixes the point where, the time when, and the numerical force
with which the battle is to be fought. By this triple determination it
has therefore a very essential influence on the issue of the combat. If
tactics has fought the battle, if the result is over, let it be victory
or defeat, Strategy makes such use of it as can be made in accordance
with the great object of the War. This object is naturally often a very
distant one, seldom does it lie quite close at hand. A series of other
objects subordinate themselves to it as means. These objects, which
are at the same time means to a higher purpose, may be practically of
various kinds; even the ultimate aim of the whole War may be a different
one in every case. We shall make ourselves acquainted with these things
according as we come to know the separate objects which they come, in
contact with; and it is not our intention here to embrace the whole
subject by a complete enumeration of them, even if that were possible.
We therefore let the employment of the battle stand over for the
present.

Even those things through which Strategy has an influence on the issue of the combat, inasmuch as it establishes the same, to a certain extent decrees them, are not so simple that they can be embraced in one single view. For as Strategy appoints time, place and force, it can do so in practice in many ways, each of which influences in a different manner the result of the combat as well as its consequences. Therefore we shall only get acquainted with this also by degrees, that is, through the subjects which more closely determine the application.

If we strip the combat of all modifications which it may undergo according to its immediate purpose and the circumstances from which it proceeds, lastly if we set aside the valour of the troops, because that is a given quantity, then there remains only the bare conception of the combat, that is a combat without form, in which we distinguish nothing but the number of the combatants.

This number will therefore determine victory. Now from the number of things above deducted to get to this point, it is shown that the superiority in numbers in a battle is only one of the factors employed to produce victory that therefore so far from having with the superiority in number obtained all, or even only the principal thing, we have perhaps got very little by it, according as the other circumstances which co-operate happen to vary.

But this superiority has degrees, it may be imagined as twofold, threefold or fourfold, and every one sees, that by increasing in this way, it must (at last) overpower everything else.

In such an aspect we grant, that the superiority in numbers is the most important factor in the result of a combat, only it must be sufficiently great to be a counterpoise to all the other co-operating circumstances. The direct result of this is, that the greatest possible number of troops should be brought into action at the decisive point.

Whether the troops thus brought are sufficient or not, we have then done in this respect all that our means allowed. This is the first principle in Strategy, therefore in general as now stated, it is just as well suited for Greeks and Persians, or for Englishmen and Mahrattas, as for French and Germans. But we shall take a glance at our relations in Europe, as respects War, in order to arrive at some more definite idea on this subject.

Here we find Armies much more alike in equipment, organisation, and practical skill of every kind. There only remains a difference in the military virtue of Armies, and in the talent of Generals which may fluctuate with time from side to side. If we go through the military history of modern Europe, we find no example of a Marathon.

Frederick the Great beat 80,000 Austrians at Leuthen with about 30,000 men, and at Rosbach with 25,000 some 50,000 allies; these are however the only instances of victories gained against an enemy double, or more than double in numbers. Charles XII, in the battle of Narva, we cannot well quote, for the Russians were at that time hardly to be regarded as Europeans, also the principal circumstances, even of the battle, are too little known. Buonaparte had at Dresden 120,000 against 220,000, therefore not the double. At Kollin, Frederick the Great did not succeed, with 30,000 against 50,000 Austrians, neither did Buonaparte in the desperate battle of Leipsic, where he was 160,000 strong, against 280,000.

From this we may infer, that it is very difficult in the present state of Europe, for the most talented General to gain a victory over an enemy double his strength. Now if we see double numbers prove such a weight in the scale against the greatest Generals, we may be sure, that in ordinary cases, in small as well as great combats, an important superiority of numbers, but which need not be over two to one, will be sufficient to ensure the victory, however disadvantageous other circumstances may be. Certainly, we may imagine a defile which even tenfold would not suffice to force, but in such a case it can be no question of a battle at all.

We think, therefore, that under our conditions, as well as in all
similar ones, the superiority at the decisive point is a matter of
capital importance, and that this subject, in the generality of cases,
is decidedly the most important of all. The strength at the decisive
point depends on the absolute strength of the Army, and on skill in
making use of it.

The first rule is therefore to enter the field with an Army as strong
as possible. This sounds very like a commonplace, but still it is really
not so.

In order to show that for a long time the strength of forces was by no
means regarded as a chief point, we need only observe, that in most,
and even in the most detailed histories of the Wars in the eighteenth
century, the strength of the Armies is either not given at all, or
only incidentally, and in no case is any special value laid upon it.
Tempelhof in his history of the Seven Years' War is the earliest writer
who gives it regularly, but at the same time he does it only very
superficially.

Even Massenbach, in his manifold critical observations on the Prussian
campaigns of 1793-94 in the Vosges, talks a great deal about hills and
valleys, roads and footpaths, but does not say a syllable about mutual
strength.

Another proof lies in a wonderful notion which haunted the heads of many
critical historians, according to which there was a certain size of an
Army which was the best, a normal strength, beyond which the forces in
excess were burdensome rather than serviceable.(*)

> (*) Tempelhof and Montalembert are the first we recollect as
> examples--the first in a passage of his first part, page
> 148; the other in his correspondence relative to the plan of
> operations of the Russians in 1759.

Lastly, there are a number of instances to be found, in which all the
available forces were not really brought into the battle,(*) or into the
War, because the superiority of numbers was not considered to have that
importance which in the nature of things belongs to it.

(*) The Prussians at Jena, 1806. Wellington at Waterloo.

If we are thoroughly penetrated with the conviction that with a
considerable superiority of numbers everything possible is to be
effected, then it cannot fail that this clear conviction reacts on the
preparations for the War, so as to make us appear in the field with
as many troops as possible, and either to give us ourselves the
superiority, or at least to guard against the enemy obtaining it. So
much for what concerns the absolute force with which the War is to be
conducted.

The measure of this absolute force is determined by the Government; and
although with this determination the real action of War commences, and
it forms an essential part of the Strategy of the War, still in most
cases the General who is to command these forces in the War must regard
their absolute strength as a given quantity, whether it be that he has
had no voice in fixing it, or that circumstances prevented a sufficient
expansion being given to it.

There remains nothing, therefore, where an absolute superiority is not
attainable, but to produce a relative one at the decisive point, by
making skilful use of what we have.

The calculation of space and time appears as the most essential thing to
this end--and this has caused that subject to be regarded as one which

embraces nearly the whole art of using military forces. Indeed, some
have gone so far as to ascribe to great strategists and tacticians a
mental organ peculiarly adapted to this point.

But the calculation of time and space, although it lies universally at
the foundation of Strategy, and is to a certain extent its daily bread,
is still neither the most difficult, nor the most decisive one.

If we take an unprejudiced glance at military history, we shall find
that the instances in which mistakes in such a calculation have proved
the cause of serious losses are very rare, at least in Strategy. But if
the conception of a skilful combination of time and space is fully to
account for every instance of a resolute and active Commander beating
several separate opponents with one and the same army (Frederick
the Great, Buonaparte), then we perplex ourselves unnecessarily with
conventional language. For the sake of clearness and the profitable use
of conceptions, it is necessary that things should always be called by
their right names.

The right appreciation of their opponents (Daun, Schwartzenberg), the
audacity to leave for a short space of time only before
them, energy in forced marches, boldness in sudden attacks, the
intensified activity which great souls acquire in the moment of danger,
these are the grounds of such victories; and what have these to do with
the ability to make an exact calculation of two such simple things as
time and space?

But even this ricochetting play of forces, "when the victories at
Rosbach and Montmirail give the impulse to victories at Leuthen and
Montereau," to which great Generals on the defensive have often trusted,
is still, if we would be clear and exact, only a rare occurrence in
history.

Much more frequently the relative superiority--that is, the skilful
assemblage of superior forces at the decisive point--has its foundation
in the right appreciation of those points, in the judicious direction
which by that means has been given to the forces from the very first,
and in the resolution required to sacrifice the unimportant to the
advantage of the important--that is, to keep the forces concentrated in
an overpowering mass. In this, Frederick the Great and Buonaparte are
particularly characteristic.

We think we have now allotted to the superiority in numbers the
importance which belongs to it; it is to be regarded as the fundamental
idea, always to be aimed at before all and as far as possible.

But to regard it on this account as a necessary condition of victory
would be a complete misconception of our exposition; in the conclusion
to be drawn from it there lies nothing more than the value which should
attach to numerical strength in the combat. If that strength is made as
great as possible, then the maxim is satisfied; a review of the total
relations must then decide whether or not the combat is to be avoided
for want of sufficient force.(*)

> (*) Owing to our freedom from invasion, and to the condition
> which arise in our Colonial Wars, we have not yet, in
> England, arrived at a correct appreciation of the value of
> superior numbers in War, and still adhere to the idea of an
> Army just "big enough," which Clausewitz has so unsparingly
> ridiculed. (EDITOR.)

CHAPTER IX. THE SURPRISE

FROM the subject of the foregoing chapter, the general endeavour to
attain a relative superiority, there follows another endeavour which
must consequently be just as general in its nature: this is the

On War.txt
SURPRISE of the enemy. It lies more or less at the foundation of all
undertakings, for without it the preponderance at the decisive point is
not properly conceivable.

The surprise is, therefore, not only the means to the attainment of
numerical superiority; but it is also to be regarded as a substantive
principle in itself, on account of its moral effect. When it is
successful in a high degree, confusion and broken courage in the enemy's
ranks are the consequences; and of the degree to which these multiply
a success, there are examples enough, great and small. We are not now
speaking of the particular surprise which belongs to the attack, but of
the endeavour by measures generally, and especially by the distribution
of forces, to surprise the enemy, which can be imagined just as well in
the defensive, and which in the tactical defence particularly is a chief
point.

We say, surprise lies at the foundation of all undertakings without
exception, only in very different degrees according to the nature of the
undertaking and other circumstances.

This difference, indeed, originates in the properties or peculiarities
of the Army and its Commander, in those even of the Government.

Secrecy and rapidity are the two factors in this product and these
suppose in the Government and the Commander-in-Chief great energy, and
on the part of the Army a high sense of military duty. With effeminacy
and loose principles it is in vain to calculate upon a surprise. But so
general, indeed so indispensable, as is this endeavour, and true as it
is that it is never wholly unproductive of effect, still it is not
the less true that it seldom succeeds to a REMARKABLE degree, and this
follows from the nature of the idea itself. We should form an erroneous
conception if we believed that by this means chiefly there is much to be
attained in War. In idea it promises a great deal; in the execution it
generally sticks fast by the friction of the whole machine.

In tactics the surprise is much more at home, for the very natural
reason that all times and spaces are on a smaller scale. It will,
therefore, in Strategy be the more feasible in proportion as the
measures lie nearer to the province of tactics, and more difficult the
higher up they lie towards the province of policy.

The preparations for a War usually occupy several months; the assembly
of an Army at its principal positions requires generally the formation
of depots and magazines, and long marches, the object of which can be
guessed soon enough.

It therefore rarely happens that one State surprises another by a
War, or by the direction which it gives the mass of its forces. In the
seventeenth and eighteenth centuries, when War turned very much upon
sieges, it was a frequent aim, and quite a peculiar and important
chapter in the Art of War, to invest a strong place unexpectedly, but
even that only rarely succeeded.(*)

    (*) Railways, steamships, and telegraphs have, however,
    enormously modified the relative importance and
    practicability of surprise. (EDITOR.)

On the other hand, with things which can be done in a day or two, a
surprise is much more conceivable, and, therefore, also it is often not
difficult thus to gain a march upon the enemy, and thereby a position, a
point of country, a road, &c. But it is evident that what surprise gains
in this way in easy execution, it loses in the efficacy, as the greater
the efficacy the greater always the difficulty of execution. Whoever
thinks that with such surprises on a small scale, he may connect great
results--as, for example, the gain of a battle, the capture of an
important magazine--believes in something which it is certainly very
possible to imagine, but for which there is no warrant in history; for
there are upon the whole very few instances where anything great has

resulted from such surprises; from which we may justly conclude that
inherent difficulties lie in the way of their success.

Certainly, whoever would consult history on such points must not depend
on sundry battle steeds of historical critics, on their wise dicta and
self-complacent terminology, but look at facts with his own eyes. There
is, for instance, a certain day in the campaign in Silesia, 1761, which,
in this respect, has attained a kind of notoriety. It is the 22nd July,
on which Frederick the Great gained on Laudon the march to Nossen, near
Neisse, by which, as is said, the junction of the Austrian and Russian
armies in Upper Silesia became impossible, and, therefore, a period of
four weeks was gained by the King. Whoever reads over this occurrence
carefully in the principal histories,(*) and considers it impartially,
will, in the march of the 22nd July, never find this importance; and
generally in the whole of the fashionable logic on this subject, he will
see nothing but contradictions; but in the proceedings of Laudon, in
this renowned period of manoeuvres, much that is unaccountable. How
could one, with a thirst for truth, and clear conviction, accept such
historical evidence?

(*) Tempelhof, The Veteran, Frederick the Great. Compare
also (Clausewitz) "Hinterlassene Werke," vol. x., p. 158.

When we promise ourselves great effects in a campaign from the principle
of surprising, we think upon great activity, rapid resolutions, and
forced marches, as the means of producing them; but that these things,
even when forthcoming in a very high degree, will not always produce the
desired effect, we see in examples given by Generals, who may be allowed
to have had the greatest talent in the use of these means, Frederick the
Great and Buonaparte. The first when he left Dresden so suddenly in
July 1760, and falling upon Lascy, then turned against Dresden, gained
nothing by the whole of that intermezzo, but rather placed his affairs
in a condition notably worse, as the fortress Glatz fell in the
meantime.

In 1813, Buonaparte turned suddenly from Dresden twice against Bluecher,
to say nothing of his incursion into Bohemia from Upper Lusatia, and
both times without in the least attaining his object. They were blows in
the air which only cost him time and force, and might have placed him in
a dangerous position in Dresden.

Therefore, even in this field, a surprise does not necessarily meet with
great success through the mere activity, energy, and resolution of the
Commander; it must be favoured by other circumstances. But we by
no means deny that there can be success; we only connect with it a
necessity of favourable circumstances, which, certainly do not occur
very frequently, and which the Commander can seldom bring about himself.

Just those two Generals afford each a striking illustration of this. We
take first Buonaparte in his famous enterprise against Bluecher's
Army in February 1814, when it was separated from the Grand Army, and
descending the Marne. It would not be easy to find a two days' march to
surprise the enemy productive of greater results than this; Bluecher's
Army, extended over a distance of three days' march, was beaten in
detail, and suffered a loss nearly equal to that of defeat in a great
battle. This was completely the effect of a surprise, for if Bluecher
had thought of such a near possibility of an attack from Buonaparte(*)
he would have organised his march quite differently. To this mistake of
Bluecher's the result is to be attributed. Buonaparte did not know all
these circumstances, and so there was a piece of good fortune that mixed
itself up in his favour.

(*) Bluecher believed his march to be covered by Pahlen's
Cossacks, but these had been withdrawn without warning to
him by the Grand Army Headquarters under Schwartzenberg.

It is the same with the battle of Liegnitz, 1760. Frederick the Great
gained this fine victory through altering during the night a position

which he had just before taken up. Laudon was through this completely
surprised, and lost 70 pieces of artillery and 10,000 men. Although
Frederick the Great had at this time adopted the principle of moving
backwards and forwards in order to make a battle impossible, or at least
to disconcert the enemy's plans, still the alteration of position on the
night of the 14-15 was not made exactly with that intention, but as the
King himself says, because the position of the 14th did not please
him. Here, therefore, also chance was hard at work; without this happy
conjunction of the attack and the change of position in the night, and
the difficult nature of the country, the result would not have been the
same.

Also in the higher and highest province of Strategy there are some
instances of surprises fruitful in results. We shall only cite the
brilliant marches of the Great Elector against the Swedes from Franconia
to Pomerania and from the Mark (Brandenburg) to the Pregel in 1757, and
the celebrated passage of the Alps by Buonaparte, 1800. In the latter
case an Army gave up its whole theatre of war by a capitulation, and in
1757 another Army was very near giving up its theatre of war and itself
as well. Lastly, as an instance of a War wholly unexpected, we may
bring forward the invasion of Silesia by Frederick the Great. Great and
powerful are here the results everywhere, but such events are not common
in history if we do not confuse with them cases in which a State, for
want of activity and energy (Saxony 1756, and Russia, 1812), has not
completed its preparations in time.

Now there still remains an observation which concerns the essence of the
thing. A surprise can only be effected by that party which gives the law
to the other; and he who is in the right gives the law. If we surprise
the adversary by a wrong measure, then instead of reaping good results,
we may have to bear a sound blow in return; in any case the adversary
need not trouble himself much about our surprise, he has in our mistake
the means of turning off the evil. As the offensive includes in itself
much more positive action than the defensive, so the surprise is
certainly more in its place with the assailant, but by no means
invariably, as we shall hereafter see. Mutual surprises by the offensive
and defensive may therefore meet, and then that one will have the
advantage who has hit the nail on the head the best.

So should it be, but practical life does not keep to this line so
exactly, and that for a very simple reason. The moral effects which
attend a surprise often convert the worst case into a good one for
the side they favour, and do not allow the other to make any regular
determination. We have here in view more than anywhere else not only the
chief Commander, but each single one, because a surprise has the effect
in particular of greatly loosening unity, so that the individuality of
each separate leader easily comes to light.

Much depends here on the general relation in which the two parties stand
to each other. If the one side through a general moral superiority can
intimidate and outdo the other, then he can make use of the surprise
with more success, and even reap good fruit where properly he should
come to ruin.

CHAPTER X. STRATAGEM

STRATAGEM implies a concealed intention, and therefore is opposed to
straightforward dealing, in the same way as wit is the opposite
of direct proof. It has therefore nothing in common with means of
persuasion, of self-interest, of force, but a great deal to do with
deceit, because that likewise conceals its object. It is itself a deceit
as well when it is done, but still it differs from what is commonly
called deceit, in this respect that there is no direct breach of word.
The deceiver by stratagem leaves it to the person himself whom he is
deceiving to commit the errors of understanding which at last, flowing
into ONE result, suddenly change the nature of things in his eyes.

We may therefore say, as nit is a sleight of hand with ideas and conceptions, so stratagem is a sleight of hand with actions.

At first sight it appears as if Strategy had not improperly derived its name from stratagem; and that, with all the real and apparent changes which the whole character of War has undergone since the time of the Greeks, this term still points to its real nature.

If we leave to tactics the actual delivery of the blow, the battle itself, and look upon Strategy as the art of using this means with skill, then besides the forces of the character, such as burning ambition which always presses like a spring, a strong will which hardly bends &c. &c., there seems no subjective quality so suited to guide and inspire strategic activity as stratagem. The general tendency to surprise, treated of in the foregoing chapter, points to this conclusion, for there is a degree of stratagem, be it ever so small, which lies at the foundation of every attempt to surprise.

But however much we feel a desire to see the actors in War outdo each other in hidden activity, readiness, and stratagem, still we must admit that these qualities show themselves but little in history, and have rarely been able to work their way to the surface from amongst the mass of relations and circumstances.

The explanation of this is obvious, and it is almost identical with the subject matter of the preceding chapter.

Strategy knows no other activity than the regulating of combat with the measures which relate to it. It has no concern, like ordinary life, with transactions which consist merely of words--that is, in expressions, declarations, &c. But these, which are very inexpensive, are chiefly the means with which the wily one takes in those he practises upon.

That which there is like it in War, plans and orders given merely as make-believers, false reports sent on purpose to the enemy--is usually of so little effect in the strategic field that it is only resorted to in particular cases which offer of themselves, therefore cannot be regarded as spontaneous action which emanates from the leader.

But such measures as carrying out the arrangements for a battle, so far as to impose upon the enemy, require a considerable expenditure of time and power; of course, the greater the impression to be made, the greater the expenditure in these respects. And as this is usually not given for the purpose, very few demonstrations, so-called, in Strategy, effect the object for which they are designed. In fact, it is dangerous to detach large forces for any length of time merely for a trick, because there is always the risk of its being done in vain, and then these forces are wanted at the decisive point.

The chief actor in War is always thoroughly sensible of this sober truth, and therefore he has no desire to play at tricks of agility. The bitter earnestness of necessity presses so fully into direct action that there is no room for that game. In a word, the pieces on the strategical chess-board want that mobility which is the element of stratagem and subtility.

The conclusion which we draw, is that a correct and penetrating eye is a more necessary and more useful quality for a General than craftiness, although that also does no harm if it does not exist at the expense of necessary qualities of the heart, which is only too often the case.

But the weaker the forces become which are under the command of Strategy, so much the more they become adapted for stratagem, so that to the quite feeble and little, for whom no prudence, no sagacity is any longer sufficient at the point where all art seems to forsake him, stratagem offers itself as a last resource. The more helpless his situation, the more everything presses towards one single, desperate blow, the more readily stratagem comes to the aid of his boldness. Let

loose from all further calculations, freed from all concern for the
future, boldness and stratagem intensify each other, and thus collect at
one point an infinitesimal glimmering of hope into a single ray, which
may likewise serve to kindle a flame.

CHAPTER XI. ASSEMBLY OF FORCES IN SPACE

THE best Strategy is ALWAYS TO BE VERY STRONG, first generally then at
the decisive point. Therefore, apart from the energy which creates the
Army, a work which is not always done by the General, there is no more
imperative and no simpler law for Strategy than to KEEP THE FORCES
CONCENTRATED.--No portion is to be separated from the main body unless
called away by some urgent necessity. On this maxim we stand firm, and
look upon it as a guide to be depended upon. What are the reasonable
grounds on which a detachment of forces may be made we shall learn by
degrees. Then we shall also see that this principle cannot have the same
general effects in every War, but that these are different according to
the means and end.

It seems incredible, and yet it has happened a hundred times, that
troops have been divided and separated merely through a mysterious
feeling of conventional manner, without any clear perception of the
reason.

If the concentration of the whole force is acknowledged as the norm, and
every division and separation as an exception which must be justified,
then not only will that folly be completely avoided, but also many an
erroneous ground for separating troops will be barred admission.

CHAPTER XII. ASSEMBLY OF FORCES IN TIME

WE have here to deal with a conception which in real life diffuses many
kinds of illusory light. A clear definition and development of the idea
is therefore necessary, and we hope to be allowed a short analysis.

War is the shock of two opposing forces in collision with each other,
from which it follows as a matter of course that the stronger not only
destroys the other, but carries it forward with it in its movement. This
fundamentally admits of no successive action of powers, but makes the
simultaneous application of all forces intended for the shock appear as
a primordial law of War.

So it is in reality, but only so far as the struggle resembles also in
practice a mechanical shock, but when it consists in a lasting, mutual
action of destructive forces, then we can certainly imagine a successive
action of forces. This is the case in tactics, principally because
firearms form the basis of all tactics, but also for other reasons as
well. If in a fire combat 1000 men are opposed to 500, then the gross
loss is calculated from the amount of the enemy's force and our own;
1000 men fire twice as many shots as 500, but more shots will take
effect on the 1000 than on the 500 because it is assumed that they stand
in closer order than the other. If we were to suppose the number of hits
to be double, then the losses on each side would be equal. From the 500
there would be for example 200 disabled, and out of the body of 1000
likewise the same; now if the 500 had kept another body of equal number
quite out of fire, then both sides would have 800 effective men; but
of these, on the one side there would be 500 men quite fresh, fully
supplied with ammunition, and in their full vigour; on the other
side only 800 all alike shaken in their order, in want of sufficient
ammunition and weakened in physical force. The assumption that the 1000
men merely on account of their greater number would lose twice as
many as 500 would have lost in their place, is certainly not correct;
therefore the greater loss which the side suffers that has placed the
half of its force in reserve, must be regarded as a disadvantage in that

original formation; further it must be admitted, that in the generality
of cases the 1000 men would have the advantage at the first commencement
of being able to drive their opponent out of his position and force
him to a retrograde movement; now, whether these two advantages are a
counterpoise to the disadvantage of finding ourselves with 800 men to
a certain extent disorganised by the combat, opposed to an enemy who is
not materially weaker in numbers and who has 500 quite fresh troops, is
one that cannot be decided by pursuing an analysis further, we must here
rely upon experience, and there will scarcely be an officer experienced
in War who will not in the generality of cases assign the advantage to
that side which has the fresh troops.

In this way it becomes evident how the employment of too many forces in
combat may be disadvantageous; for whatever advantages the superiority
may give in the first moment, we may have to pay dearly for in the next.

But this danger only endures as long as the disorder, the state of
confusion and weakness lasts, in a word, up to the crisis which every
combat brings with it even for the conqueror. Within the duration of
this relaxed state of exhaustion, the appearance of a proportionate
number of fresh troops is decisive.

But when this disordering effect of victory stops, and therefore only
the moral superiority remains which every victory gives, then it is no
longer possible for fresh troops to restore the combat, they would
only be carried along in the general movement; a beaten Army cannot be
brought back to victory a day after by means of a strong reserve. Here
we find ourselves at the source of a highly material difference between
tactics and strategy.

The tactical results, the results within the four corners of the battle,
and before its close, lie for the most part within the limits of that
period of disorder and weakness. But the strategic result, that is to
say, the result of the total combat, of the victories realised, let them
be small or great, lies completely (beyond) outside of that period.
It is only when the results of partial combats have bound themselves
together into an independent whole, that the strategic result appears,
but then, the state of crisis is over, the forces have resumed their
original form, and are now only weakened to the extent of those actually
destroyed (placed hors de combat).

The consequence of this difference is, that tactics can make a continued
use of forces, Strategy only a simultaneous one.(*)

> (*) See chaps. xiii., and xiv., Book III and chap. xxix.
> Book V.--TR.

If I cannot, in tactics, decide all by the first success, if I have to
fear the next moment, it follows of itself that I employ only so much of
my force for the success of the first moment as appears sufficient for
that object, and keep the rest beyond the reach of fire or conflict of
any kind, in order to be able to oppose fresh troops to fresh, or with
such to overcome those that are exhausted. But it is not so in Strategy.
Partly, as we have just shown, it has not so much reason to fear a
reaction after a success realised, because with that success the crisis
stops; partly all the forces strategically employed are not necessarily
weakened. Only so much of them as have been tactically in conflict with
the enemy's force, that is, engaged in partial combat, are weakened by
it; consequently, only so much as was unavoidably necessary, but by no
means all which was strategically in conflict with the enemy, unless
tactics has expended them unnecessarily. Corps which, on account of the
general superiority in numbers, have either been little or not at all
engaged, whose presence alone has assisted in the result, are after
the decision the same as they were before, and for new enterprises as
efficient as if they had been entirely inactive. How greatly such corps
which thus constitute our excess may contribute to the total success is
evident in itself; indeed, it is not difficult to see how they may
even diminish considerably the loss of the forces engaged in tactical,

conflict on our side.

If, therefore, in Strategy the loss does not increase with the number of
the troops employed, but is often diminished by it, and if, as a natural
consequence, the decision in our favor is, by that means, the more
certain, then it follows naturally that in Strategy we can never
employ too many forces, and consequently also that they must be applied
simultaneously to the immediate purpose.

But we must vindicate this proposition upon another ground. We have
hitherto only spoken of the combat itself; it is the real activity in
War, but men, time, and space, which appear as the elements of this
activity, must, at the same time, be kept in view, and the results of
their influence brought into consideration also.

Fatigue, exertion, and privation constitute in War a special principle
of destruction, not essentially belonging to contest, but more or less
inseparably bound up with it, and certainly one which especially belongs
to Strategy. They no doubt exist in tactics as well, and perhaps there
in the highest degree; but as the duration of the tactical acts is
shorter, therefore the small effects of exertion and privation on them
can come but little into consideration. But in Strategy on the other
hand, where time and space, are on a larger scale, their influence is
not only always very considerable, but often quite decisive. It is not
at all uncommon for a victorious Army to lose many more by sickness than
on the field of battle.

If, therefore, we look at this sphere of destruction in Strategy in
the same manner as we have considered that of fire and close combat in
tactics, then we may well imagine that everything which comes within
its vortex will, at the end of the campaign or of any other strategic
period, be reduced to a state of weakness, which makes the arrival of a
fresh force decisive. We might therefore conclude that there is a motive
in the one case as well as the other to strive for the first success
with as few forces as possible, in order to keep up this fresh force for
the last.

In order to estimate exactly this conclusion, which, in many cases
in practice, will have a great appearance of truth, we must direct our
attention to the separate ideas which it contains. In the first place,
we must not confuse the notion of reinforcement with that of fresh
unused troops. There are few campaigns at the end of which an increase
of force is not earnestly desired by the conqueror as well as the
conquered, and indeed should appear decisive; but that is not the point
here, for that increase of force could not be necessary if the force
had been so much larger at the first. But it would be contrary to all
experience to suppose that an Army coming fresh into the field is to
be esteemed higher in point of moral value than an Army already in the
field, just as a tactical reserve is more to be esteemed than a body
of troops which has already been severely handled in the fight. Just as
much as an unfortunate campaign lowers the courage and moral powers of
an Army, a successful one raises these elements in their value. In the
generality of cases, therefore, these influences are compensated, and
then there remains over and above as clear gain the habituation to War.
We should besides look more here to successful than to unsuccessful
campaigns, because when the greater probability of the latter may be
seen beforehand, without doubt forces are wanted, and, therefore, the
reserving a portion for future use is out of the question.

This point being settled, then the question is, Do the losses which a
force sustains through fatigues and privations increase in proportion to
the size of the force, as is the case in a combat? And to that we answer
"No."

The fatigues of War result in a great measure from the dangers with
which every moment of the act of War is more or less impregnated. To
encounter these dangers at all points, to proceed onwards with security
in the execution of one's plans, gives employment to a multitude of

agencies which make up the tactical and strategic service of the Army. This service is more difficult the weaker an Army is, and easier as its numerical superiority over that of the enemy increases. Who can doubt this? A campaign against a much weaker enemy will therefore cost smaller efforts than against one just as strong or stronger.

So much for the fatigues. It is somewhat different with the privations; they consist chiefly of two things, the want of food, and the want of shelter for the troops, either in quarters or in suitable camps. Both these wants will no doubt be greater in proportion as the number of men on one spot is greater. But does not the superiority in force afford also the best means of spreading out and finding more room, and therefore more means of subsistence and shelter?

If Buonaparte, in his invasion of Russia in 1812, concentrated his Army in great masses upon one single road in a manner never heard of before, and thus caused privations equally unparalleled, we must ascribe it to his maxim THAT IT IS IMPOSSIBLE TO BE TOO STRONG AT THE DECISIVE POINT. Whether in this instance he did not strain the principle too far is a question which would be out of place here; but it is certain that, if he had made a point of avoiding the distress which was by that means brought about, he had only to advance on a greater breadth of front. Room was not wanted for the purpose in Russia, and in very few cases can it be wanted. Therefore, from this no ground can be deduced to prove that the simultaneous employment of very superior forces must produce greater weakening. But now, supposing that in spite of the general relief afforded by setting apart a portion of the Army, wind and weather and the toils of war had produced a diminution even on the part which as a spare force had been reserved for later use, still we must take a comprehensive general view of the whole, and therefore ask, will this diminution of force suffice to counterbalance the gain in forces, which we, through our superiority in numbers, may be able to make in more ways than one?

But there still remains a most important point to be noticed. In a partial combat, the force required to obtain a great result can be approximately estimated without much difficulty, and, consequently, we can form an idea of what is superfluous. In Strategy this may be said to be impossible, because the strategic result has no such well-defined object and no such circumscribed limits as the tactical. Thus what can be looked upon in tactics as an excess of power, must be regarded in Strategy as a means to give expansion to success, if opportunity offers for it; with the magnitude of the success the gain in force increases at the same time, and in this way the superiority of numbers may soon reach a point which the most careful economy of forces could never have attained.

By means of his enormous numerical superiority, Buonaparte was enabled to reach Moscow in 1812, and to take that central capital. Had he by means of this superiority succeeded in completely defeating the Russian Army, he would, in all probability, have concluded a peace in Moscow which in any other way was much less attainable. This example is used to explain the idea, not to prove it, which would require a circumstantial demonstration, for which this is not the place.(*)

(*) Compare Book VII., second edition, p. 56.

All these reflections bear merely upon the idea of a successive employment of forces, and not upon the conception of a reserve properly so called, which they, no doubt, come in contact with throughout, but which, as we shall see in the following chapter, is connected with some other considerations.

What we desire to establish here is, that if in tactics the military force through the mere duration of actual employment suffers a diminution of power, if time, therefore, appears as a factor in the result, this is not the case in Strategy in a material degree. The destructive effects which are also produced upon the forces in Strategy by time, are partly diminished through their mass, partly made good in other ways, and, therefore, in Strategy it cannot be an object to make time an ally on its own account by bringing troops successively into action.

We say on "its own account," for the influence which time, on account of other circumstances which it brings about but which are different from itself can have, indeed must necessarily have, for one of the two parties, is quite another thing, is anything but indifferent or unimportant, and will be the subject of consideration hereafter.

The rule which we have been seeking to set forth is, therefore, that all forces which are available and destined for a strategic object should be SIMULTANEOUSLY applied to it; and this application will be so much the more complete the more everything is compressed into one act and into one movement.

But still there is in Strategy a renewal of effort and a persistent action which, as a chief means towards the ultimate success, is more particularly not to be overlooked, it is the CONTINUAL DEVELOPMENT OF NEW FORCES. This is also the subject of another chapter, and we only refer to it here in order to prevent the reader from having something in view of which we have not been speaking.

We now turn to a subject very closely connected with our present considerations, which must be settled before full light can be thrown on the whole, we mean the STRATEGIC RESERVE.

CHAPTER XIII. STRATEGIC RESERVE

A RESERVE has two objects which are very distinct from each other, namely, first, the prolongation and renewal of the combat, and secondly, for use in case of unforeseen events. The first object implies the utility of a successive application of forces, and on that account cannot occur in Strategy. Cases in which a corps is sent to succour a point which is supposed to be about to fall are plainly to be placed in the category of the second object, as the resistance which has to be offered here could not have been sufficiently foreseen. But a corps which is destined expressly to prolong the combat, and with that object in view is placed in rear, would be only a corps placed out of reach of fire, but under the command and at the disposition of the General Commanding in the action, and accordingly would be a tactical and not a strategic reserve.

But the necessity for a force ready for unforeseen events may also take place in Strategy, and consequently there may also be a strategic reserve, but only where unforeseen events are imaginable. In tactics, where the enemy's measures are generally first ascertained by direct sight, and where they may be concealed by every wood, every fold of undulating ground, we must naturally always be alive, more or less, to the possibility of unforeseen events, in order to strengthen, subsequently, those points which appear too weak, and, in fact, to modify generally the disposition of our troops, so as to make it correspond better to that of the enemy.

Such cases must also happen in Strategy, because the strategic act is directly linked to the tactical. In Strategy also many a measure is first adopted in consequence of what is actually seen, or in consequence of uncertain reports arriving from day to day, or even from hour to hour, and lastly, from the actual results of the combats it is, therefore, an essential condition of strategic command that, according to the degree of uncertainty, forces must be kept in reserve against future contingencies.

In the defensive generally, but particularly in the defence of certain obstacles of ground, like rivers, hills, &c., such contingencies, as is well known, happen constantly.

But this uncertainty diminishes in proportion as the strategic activity has less of the tactical character, and ceases almost altogether in those regions where it borders on politics.

The direction in which the enemy leads his columns to the combat can be perceived by actual sight only; where he intends to pass a river is learnt from a few preparations which are made shortly before; the line by which he proposes to invade our country is usually announced by all the newspapers before a pistol shot has been fired. The greater the nature of the measure the less it will take the enemy by surprise. Time and space are so considerable, the circumstances out of which the action proceeds so public and little susceptible of alteration, that the coming event is either made known in good time, or can be discovered with reasonable certainty.

On the other hand the use of a reserve in this province of Strategy, even if one were available, will always be less efficacious the more the measure has a tendency towards being one of a general nature.

We have seen that the decision of a partial combat is nothing in itself, but that all partial combats only find their complete solution in the decision of the total combat.

But even this decision of the total combat has only a relative meaning of many different gradations, according as the force over which the victory has been gained forms a more or less great and important part of the whole. The lost battle of a corps may be repaired by the victory of the Army. Even the lost battle of an Army may not only be counterbalanced by the gain of a more important one, but converted into a fortunate event (the two days of Kulm, August 29 and 30, 1813(*)). No one can doubt this; but it is just as clear that the weight of each victory (the successful issue of each total combat) is so much the more substantial the more important the part conquered, and that therefore the possibility of repairing the loss by subsequent events diminishes in the same proportion. In another place we shall have to examine this more in detail; it suffices for the present to have drawn attention to the indubitable existence of this progression.

> (*) Refers to the destruction of Vandamme's column, which
> had been sent unsupported to intercept the retreat of the
> Austrians and Prussians from Dresden--but was forgotten by
> Napoleon.--EDITOR.

If we now add lastly to these two considerations the third, which is, that if the persistent use of forces in tactics always shifts the great result to the end of the whole act, law of the simultaneous use of the forces in Strategy, on the contrary, lets the principal result (which need not be the final one) take place almost always at the commencement of the great (or whole) act, then in these three results we have grounds sufficient to find strategic reserves always more superfluous, always more useless, always more dangerous, the more general their destination.

The point where the idea of a strategic reserve begins to become inconsistent is not difficult to determine: it lies in the SUPREME DECISION. Employment must be given to all the forces within the space of the supreme decision, and every reserve (active force available) which is only intended for use after that decision is opposed to common sense.

If, therefore, tactics has in its reserves the means of not only meeting unforeseen dispositions on the part of the enemy, but also of repairing that which never can be foreseen, the result of the combat, should that be unfortunate; Strategy on the other hand must, at least as far as relates to the capital result, renounce the use of these means. As A rule, it can only repair the losses sustained at one point by advantages gained at another, in a few cases by moving troops from one point to another; the idea of preparing for such reverses by placing forces in reserve beforehand, can never be entertained in Strategy.

We have pointed out as an absurdity the idea of a strategic reserve
which is not to co-operate in the capital result, and as it is so beyond
a doubt, we should not have been led into such an analysis as we have
made in these two chapters, were it not that, in the disguise of
other ideas, it looks like something better, and frequently makes its
appearance. One person sees in it the acme of strategic sagacity and
foresight; another rejects it, and with it the idea of any reserve,
consequently even of a tactical one. This confusion of ideas is
transferred to real life, and if we would see a memorable instance of
it we have only to call to mind that Prussia in 1806 left a reserve
of 20,000 men cantoned in the Mark, under Prince Eugene of Wurtemberg,
which could not possibly reach the Saale in time to be of any use, and
that another force Of 25,000 men belonging to this power remained
in East and South Prussia, destined only to be put on a war-footing
afterwards as a reserve.

After these examples we cannot be accused of having been fighting with
windmills.

CHAPTER XIV. ECONOMY OF FORCES

THE road of reason, as we have said, seldom allows itself to be reduced
to a mathematical line by principles and opinions. There remains always
a certain margin. But it is the same in all the practical arts of life.
For the lines of beauty there are no abscissae and ordinates; circles
and ellipses are not described by means of their algebraical formulae.
The actor in War therefore soon finds he must trust himself to the
delicate tact of judgment which, founded on natural quickness of
perception, and educated by reflection, almost unconsciously seizes upon
the right; he soon finds that at one time he must simplify the law (by
reducing it) to some prominent characteristic points which form his
rules; that at another the adopted method must become the staff on which
he leans.

As one of these simplified characteristic points as a mental appliance,
we look upon the principle of watching continually over the co-operation
of all forces, or in other words, of keeping constantly in view that
no part of them should ever be idle. Whoever has forces where the enemy
does not give them sufficient employment, whoever has part of his forces
on the march--that is, allows them to lie dead--while the enemy's are
fighting, he is a bad manager of his forces. In this sense there is
a waste of forces, which is even worse than their employment to no
purpose. If there must be action, then the first point is that all parts
act, because the most purposeless activity still keeps employed and
destroys a portion of the enemy's force, whilst troops completely
inactive are for the moment quite neutralised. Unmistakably this idea is
bound up with the principles contained in the last three chapters, it
is the same truth, but seen from a somewhat more comprehensive point of
view and condensed into a single conception.

CHAPTER XV. GEOMETRICAL ELEMENT

THE length to which the geometrical element or form in the disposition
of military force in War can become a predominant principle, we see in
the art of fortification, where geometry looks after the great and
the little. Also in tactics it plays a great part. It is the basis of
elementary tactics, or of the theory of moving troops; but in field
fortification, as well as in the theory of positions, and of their
attack, its angles and lines rule like law givers who have to decide the
contest. Many things here were at one time misapplied, and others were
mere fribbles; still, however, in the tactics of the present day, in
which in every combat the aim is to surround the enemy, the geometrical

element has attained anew a great importance in a very simple, but
constantly recurring application. Nevertheless, in tactics, where all is
more movable, where the moral forces, individual traits, and chance are
more influential than in a war of sieges, the geometrical element can
never attain to the same degree of supremacy as in the latter. But less
still is its influence in Strategy; certainly here, also, form in the
disposition of troops, the shape of countries and states is of
great importance; but the geometrical element is not decisive, as in
fortification, and not nearly so important as in tactics.--The manner
in which this influence exhibits itself, can only be shown by degrees at
those places where it makes its appearance, and deserves notice. Here we
wish more to direct attention to the difference which there is between
tactics and Strategy in relation to it.

In tactics time and space quickly dwindle to their absolute minimum.
If a body of troops is attacked in flank and rear by the enemy, it soon
gets to a point where retreat no longer remains; such a position is
very close to an absolute impossibility of continuing the fight; it must
therefore extricate itself from it, or avoid getting into it. This gives
to all combinations aiming at this from the first commencement a great
efficiency, which chiefly consists in the disquietude which it causes
the enemy as to consequences. This is why the geometrical disposition of
the forces is such an important factor in the tactical product.

In Strategy this is only faintly reflected, on account of the greater
space and time. We do not fire from one theatre of war upon another; and
often weeks and months must pass before a strategic movement designed to
surround the enemy can be executed. Further, the distances are so great
that the probability of hitting the right point at last, even with the
best arrangements, is but small.

In Strategy therefore the scope for such combinations, that is for those
resting on the geometrical element, is much smaller, and for the same
reason the effect of an advantage once actually gained at any point
is much greater. Such advantage has time to bring all its effects to
maturity before it is disturbed, or quite neutralised therein, by any
counteracting apprehensions. We therefore do not hesitate to regard as
an established truth, that in Strategy more depends on the number and
the magnitude of the victorious combats, than on the form of the great
lines by which they are connected.

A view just the reverse has been a favourite theme of modern theory,
because a greater importance was supposed to be thus given to Strategy,
and, as the higher functions of the mind were seen in Strategy, it was
thought by that means to ennoble War, and, as it was said--through a new
substitution of ideas--to make it more scientific. We hold it to be
one of the principal uses of a complete theory openly to expose such
vagaries, and as the geometrical element is the fundamental idea from
which theory usually proceeds, therefore we have expressly brought out
this point in strong relief.

CHAPTER XVI. ON THE SUSPENSION OF THE ACT IN WARFARE

IF one considers War as an act of mutual destruction, we must of
necessity imagine both parties as making some progress; but at the same
time, as regards the existing moment, we must almost as necessarily
suppose the one party in a state of expectation, and only the other
actually advancing, for circumstances can never be actually the same on
both sides, or continue so. In time a change must ensue, from which it
follows that the present moment is more favourable to one side than the
other. Now if we suppose that both commanders have a full knowledge of
this circumstance, then the one has a motive for action, which at the
same time is a motive for the other to wait; therefore, according to
this it cannot be for the interest of both at the same time to advance,
nor can waiting be for the interest of both at the same time. This
opposition of interest as regards the object is not deduced here from

the principle of general polarity, and therefore is not in opposition to
the argument in the fifth chapter of the second book; it depends on
the fact that here in reality the same thing is at once an incentive
or motive to both commanders, namely the probability of improving or
impairing their position by future action.

But even if we suppose the possibility of a perfect equality of
circumstances in this respect, or if we take into account that through
imperfect knowledge of their mutual position such an equality may appear
to the two Commanders to subsist, still the difference of political
objects does away with this possibility of suspension. One of the
parties must of necessity be assumed politically to be the aggressor,
because no war could take place from defensive intentions on both
sides. But the aggressor has the positive object, the defender merely a
negative one. To the first then belongs the positive action, for it is
only by that means that he can attain the positive object; therefore,
in cases where both parties are in precisely similar circumstances, the
aggressor is called upon to act by virtue of his positive object.

Therefore, from this point of view, a suspension in the act of Warfare,
strictly speaking, is in contradiction with the nature of the thing;
because two Armies, being two incompatible elements, should destroy one
another unremittingly, just as fire and water can never put themselves
in equilibrium, but act and react upon one another, until one quite
disappears. What would be said of two wrestlers who remained clasped
round each other for hours without making a movement. Action in War,
therefore, like that of a clock which is wound up, should go on running
down in regular motion.--But wild as is the nature of War it still wears
the chains of human weakness, and the contradiction we see here, viz.,
that man seeks and creates dangers which he fears at the same time will
astonish no one.

If we cast a glance at military history in general, we find so much the
opposite of an incessant advance towards the aim, that STANDING STILL
and DOING NOTHING is quite plainly the NORMAL CONDITION of an Army in
the midst of War, ACTING, the EXCEPTION. This must almost raise a doubt
as to the correctness of our conception. But if military history
leads to this conclusion when viewed in the mass the latest series of
campaigns redeems our position. The War of the French Revolution shows
too plainly its reality, and only proves too clearly its necessity. In
these operations, and especially in the campaigns of Buonaparte, the
conduct of War attained to that unlimited degree of energy which we have
represented as the natural law of the element. This degree is therefore
possible, and if it is possible then it is necessary.

How could any one in fact justify in the eyes of reason the expenditure
of forces in War, if acting was not the object? The baker only heats
his oven if he has bread to put into it; the horse is only yoked to the
carriage if we mean to drive; why then make the enormous effort of a War
if we look for nothing else by it but like efforts on the part of the
enemy?

So much in justification of the general principle; now as to its
modifications, as far as they lie in the nature of the thing and are
independent of special cases.

There are three causes to be noticed here, which appear as innate
counterpoises and prevent the over-rapid or uncontrollable movement of
the wheel-work.

The first, which produces a constant tendency to delay, and is thereby
a retarding principle, is the natural timidity and want of resolution
in the human mind, a kind of inertia in the moral world, but which is
produced not by attractive, but by repellent forces, that is to say, by
dread of danger and responsibility.

In the burning element of War, ordinary natures appear to become
heavier; the impulsion given must therefore be stronger and more

frequently repeated if the motion is to be a continuous one. The
mere idea of the object for which arms have been taken up is seldom
sufficient to overcome this resistant force, and if a warlike
enterprising spirit is not at the head, who feels himself in War in his
natural element, as much as a fish in the ocean, or if there is not the
pressure from above of some great responsibility, then standing still
will be the order of the day, and progress will be the exception.

The second cause is the imperfection of human perception and judgment,
which is greater in War than anywhere, because a person hardly knows
exactly his own position from one moment to another, and can only
conjecture on slight grounds that of the enemy, which is purposely
concealed; this often gives rise to the case of both parties looking
upon one and the same object as advantageous for them, while in reality
the interest of one must preponderate; thus then each may think he acts
wisely by waiting another moment, as we have already said in the fifth
chapter of the second book.

The third cause which catches hold, like a ratchet wheel in machinery,
from time to time producing a complete standstill, is the greater
strength of the defensive form. A may feel too weak to attack B, from
which it does not follow that B is strong enough for an attack on A. The
addition of strength, which the defensive gives is not merely lost
by assuming the offensive, but also passes to the enemy just as,
figuratively expressed, the difference of a + b and a - b is equal to
2b. Therefore it may so happen that both parties, at one and the same
time, not only feel themselves too weak to attack, but also are so in
reality.

Thus even in the midst of the act of War itself, anxious sagacity and
the apprehension of too great danger find vantage ground, by means of
which they can exert their power, and tame the elementary impetuosity of
War.

However, at the same time these causes without an exaggeration of their
effect, would hardly explain the long states of inactivity which took
place in military operations, in former times, in wars undertaken about
interests of no great importance, and in which inactivity consumed
nine-tenths of the time that the troops remained under arms. This
feature in these Wars, is to be traced principally to the influence
which the demands of the one party, and the condition, and feeling of
the other, exercised over the conduct of the operations, as has been
already observed in the chapter on the essence and object of War.

These things may obtain such a preponderating influence as to make of
War a half-and-half affair. A War is often nothing more than an armed
neutrality, or a menacing attitude to support negotiations or an attempt
to gain some small advantage by small exertions, and then to wait the
tide of circumstances, or a disagreeable treaty obligation, which is
fulfilled in the most niggardly way possible.

In all these cases in which the impulse given by interest is slight,
and the principle of hostility feeble, in which there is no desire to
do much, and also not much to dread from the enemy; in short, where no
powerful motives press and drive, cabinets will not risk much in the
game; hence this tame mode of carrying on War, in which the hostile
spirit of real War is laid in irons.

The more War becomes in this manner devitalised so much the more its
theory becomes destitute of the necessary firm pivots and buttresses for
its reasoning; the necessary is constantly diminishing, the accidental
constantly increasing.

Nevertheless in this kind of warfare, there is also a certain
shrewdness, indeed, its action is perhaps more diversified, and more
extensive than in the other. Hazard played with realeaux of gold seems
changed into a game of commerce with groschen. And on this field, where
the conduct of War spins out the time with a number of small flourishes,

with skirmishes at outposts, half in earnest half in jest, with long
dispositions which end in nothing with positions and marches, which
afterwards are designated as skilful only because their infinitesimally
small causes are lost, and common sense can make nothing of them, here
on this very field many theorists find the real Art of War at home: in
these feints, parades, half and quarter thrusts of former wars, they
find the aim of all theory, the supremacy of mind over matter, and
modern wars appear to them mere savage fisticuffs, from which nothing
is to be learnt, and which must be regarded as mere retrograde steps
towards barbarism. This opinion is as frivolous as the objects to which
it relates. Where great forces and great passions are wanting, it is
certainly easier for a practised dexterity to show its game; but is
then the command of great forces, not in itself a higher exercise of the
intelligent faculties? Is then that kind of conventional sword-exercise
not comprised in and belonging to the other mode of conducting war? Does
it not bear the same relation to it as the motions upon a ship to the
motion of the ship itself? Truly it can take place only under the tacit
condition that the adversary does no better. And can we tell, how long
he may choose to respect those conditions? Has not then the French
Revolution fallen upon us in the midst of the fancied security of our
old system of war, and driven us from Chalons to Moscow? And did not
Frederick the Great in like manner surprise the Austrians reposing in
their ancient habits of war, and make their monarchy tremble? Woe to
the cabinet which, with a shilly-shally policy, and a routine-ridden
military system, meets with an adversary who, like the rude element,
knows no other law than that of his intrinsic force. Every deficiency
in energy and exertion is then a weight in the scales in favour of the
enemy; it is not so easy then to change from the fencing posture into
that of an athlete, and a slight blow is often sufficient to knock down
the whole.

The result of all the causes now adduced is, that the hostile action
of a campaign does not progress by a continuous, but by an intermittent
movement, and that, therefore, between the separate bloody acts,
there is a period of watching, during which both parties fall into the
defensive, and also that usually a higher object causes the principle of
aggression to predominate on one side, and thus leaves it in general in
an advancing position, by which then its proceedings become modified in
some degree.

CHAPTER XVII. ON THE CHARACTER OF MODERN WAR

THE attention which must be paid to the character of war as it is now
made, has a great influence upon all plans, especially on strategic
ones.

Since all methods formerly usual were upset by Buonaparte's luck and
boldness, and first-rate Powers almost wiped out at a blow; since the
Spaniards by their stubborn resistance have shown what the general
arming of a nation and insurgent measures on a great scale can effect,
in spite of weakness and porousness of individual parts; since Russia,
by the campaign of 1812 has taught us, first, that an Empire of great
dimensions is not to be conquered (which might have been easily known
before), secondly, that the probability of final success does not in all
cases diminish in the same measure as battles, capitals, and provinces
are lost (which was formerly an incontrovertible principle with all
diplomatists, and therefore made them always ready to enter at once into
some bad temporary peace), but that a nation is often strongest in
the heart of its country, if the enemy's offensive power has exhausted
itself, and with what enormous force the defensive then springs over
to the offensive; further, since Prussia (1813) has shown that sudden
efforts may add to an Army sixfold by means of the militia, and
that this militia is just as fit for service abroad as in its own
country;--since all these events have shown what an enormous factor the
heart and sentiments of a Nation may be in the product of its political
and military strength, in fine, since governments have found out all

these additional aids, it is not to be expected that they will let them lie idle in future Wars, whether it be that danger threatens their own existence, or that restless ambition drives them on.

That a War which is waged with the whole weight of the national power on each side must be organised differently in principle to those where everything is calculated according to the relations of standing Armies to each other, it is easy to perceive. Standing Armies once resembled fleets, the land force the sea force in their relations to the remainder of the State, and from that the Art of War on shore had in it something of naval tactics, which it has now quite lost.

CHAPTER XVIII. TENSION AND REST

The Dynamic Law of War

WE have seen in the sixteenth chapter of this book, how, in most campaigns, much more time used to be spent in standing still and inaction than in activity.

Now, although, as observed in the preceding chapter we see quite a different character in the present form of War, still it is certain that real action will always be interrupted more or less by long pauses; and this leads to the necessity of our examining more closely the nature of these two phases of War.

If there is a suspension of action in War, that is, if neither party wills something positive, there is rest, and consequently equilibrium, but certainly an equilibrium in the largest signification, in which not only the moral and physical war-forces, but all relations and interests, come into calculation. As soon as ever one of the two parties proposes to himself a new positive object, and commences active steps towards it, even if it is only by preparations, and as soon as the adversary opposes this, there is a tension of powers; this lasts until the decision takes place--that is, until one party either gives up his object or the other has conceded it to him.

This decision--the foundation of which lies always in the combat--combinations which are made on each side--is followed by a movement in one or other direction.

When this movement has exhausted itself, either in the difficulties which had to be mastered, in overcoming its own internal friction, or through new resistant forces prepared by the acts of the enemy, then either a state of rest takes place or a new tension with a decision, and then a new movement, in most cases in the opposite direction.

This speculative distinction between equilibrium, tension, and motion is more essential for practical action than may at first sight appear.

In a state of rest and of equilibrium a varied kind of activity may prevail on one side that results from opportunity, and does not aim at a great alteration. Such an activity may contain important combats--even pitched battles--but yet it is still of quite a different nature, and on that account generally different in its effects.

If a state of tension exists, the effects of the decision are always greater partly because a greater force of will and a greater pressure of circumstances manifest themselves therein; partly because everything has been prepared and arranged for a great movement. The decision in such cases resembles the effect of a mine well closed and tamped, whilst an event in itself perhaps just as great, in a state of rest, is more or less like a mass of powder puffed away in the open air.

At the same time, as a matter of course, the state of tension must be imagined in different degrees of intensity, and it may therefore

approach gradually by many steps towards the state of rest, so that at
the last there is a very slight difference between them.

Now the real use which we derive from these reflections is the
conclusion that every measure which is taken during a state of tension
is more important and more prolific in results than the same measure
could be in a state of equilibrium, and that this importance increases
immensely in the highest degrees of tension.

The cannonade of Valmy, September 20, 1792, decided more than the battle
of Hochkirch, October 14, 1758.

In a tract of country which the enemy abandons to us because he cannot
defend it, we can settle ourselves differently from what we should do if
the retreat of the enemy was only made with the view to a decision under
more favourable circumstances. Again, a strategic attack in course of
execution, a faulty position, a single false march, may be decisive in
its consequence; whilst in a state of equilibrium such errors must be
of a very glaring kind, even to excite the activity of the enemy in a
general way.

Most bygone Wars, as we have already said, consisted, so far as regards
the greater part of the time, in this state of equilibrium, or at least
in such short tensions with long intervals between them, and weak in
their effects, that the events to which they gave rise were seldom great
successes, often they were theatrical exhibitions, got up in honour of a
royal birthday (Hochkirch), often a mere satisfying of the honour of the
arms (Kunersdorf), or the personal vanity of the commander (Freiberg).

That a Commander should thoroughly understand these states, that he
should have the tact to act in the spirit of them, we hold to be a great
requisite, and we have had experience in the campaign of 1806 how far
it is sometimes wanting. In that tremendous tension, when everything
pressed on towards a supreme decision, and that alone with all its
consequences should have occupied the whole soul of the Commander,
measures were proposed and even partly carried out (such as the
reconnaissance towards Franconia), which at the most might have given a
kind of gentle play of oscillation within a state of equilibrium. Over
these blundering schemes and views, absorbing the activity of the Army,
the really necessary means, which could alone save, were lost sight of.

But this speculative distinction which we have made is also necessary
for our further progress in the construction of our theory, because all
that we have to say on the relation of attack and defence, and on the
completion of this double-sided act, concerns the state of the crisis in
which the forces are placed during the tension and motion, and
because all the activity which can take place during the condition of
equilibrium can only be regarded and treated as a corollary; for
that crisis is the real War and this state of equilibrium only its
reflection.

BOOK IV THE COMBAT

CHAPTER I. INTRODUCTORY

HAVING in the foregoing book examined the subjects which may be regarded
as the efficient elements of War, we shall now turn our attention to the
combat as the real activity in Warfare, which, by its physical and moral
effects, embraces sometimes more simply, sometimes in a more complex
manner, the object of the whole campaign. In this activity and in its
effects these elements must therefore, reappear.

The formation of the combat is tactical in its nature; we only glance
at it here in a general way in order to get acquainted with it in its

On War.txt

aspect as a whole. In practice the minor or more immediate objects give every combat a characteristic form; these minor objects we shall not discuss until hereafter. But these peculiarities are in comparison to the general characteristics of a combat mostly only insignificant, so that most combats are very like one another, and, therefore, in order to avoid repeating that which is general at every stage, we are compelled to look into it here, before taking up the subject of its more special application.

In the first place, therefore, we shall give in the next chapter, in a few words, the characteristics of the modern battle in its tactical course, because that lies at the foundation of our conceptions of what the battle really is.

CHAPTER II. CHARACTER OF THE MODERN BATTLE

ACCORDING to the notion we have formed of tactics and strategy, it follows, as a matter of course, that if the nature of the former is changed, that change must have an influence on the latter. If tactical facts in one case are entirely different from those in another, then the strategic, must be so also, if they are to continue consistent and reasonable. It is therefore important to characterise a general action in its modern form before we advance with the study of its employment in strategy.

What do we do now usually in a great battle? We place ourselves quietly in great masses arranged contiguous to and behind one another. We deploy relatively only a small portion of the whole, and let it wring itself out in a fire-combat which lasts for several hours, only interrupted now and again, and removed hither and thither by separate small shocks from charges with the bayonet and cavalry attacks. When this line has gradually exhausted part of its warlike ardour in this manner and there remains nothing more than the cinders, it is withdrawn(*) and replaced by another.

(*) The relief of the fighting line played a great part in the battles of the Smooth-Bore era; it was necessitated by the fouling of the muskets, physical fatigue of the men and consumption of ammunition, and was recognised as both necessary and advisable by Napoleon himself.--EDITOR.

In this manner the battle on a modified principle burns slowly away like wet powder, and if the veil of night commands it to stop, because neither party can any longer see, and neither chooses to run the risk of blind chance, then an account is taken by each side respectively of the masses remaining, which can be called still effective, that is, which have not yet quite collapsed like extinct volcanoes; account is taken of the ground gained or lost, and of how stands the security of the rear; these results with the special impressions as to bravery and cowardice, ability and stupidity, which are thought to have been observed in ourselves and in the enemy are collected into one single total impression, out of which there springs the resolution to quit the field or to renew the combat on the morrow.

This description, which is not intended as a finished picture of a modern battle, but only to give its general tone, suits for the offensive and defensive, and the special traits which are given, by the object proposed, the country, &c. &c., may be introduced into it, without materially altering the conception.

But modern battles are not so by accident; they are so because the parties find themselves nearly on a level as regards military organisation and the knowledge of the Art of War, and because the warlike element inflamed by great national interests has broken through artificial limits and now flows in its natural channel. Under these two conditions, battles will always preserve this character.

This general idea of the modern battle will be useful to us in the
sequel in more places than one, if we want to estimate the value of the
particular co-efficients of strength, country, &c. &c. It is only for
general, great, and decisive combats, and such as come near to them that
this description stands good; inferior ones have changed their character
also in the same direction but less than great ones. The proof of this
belongs to tactics; we shall, however, have an opportunity hereafter of
making this subject plainer by giving a few particulars.

CHAPTER III. THE COMBAT IN GENERAL

THE Combat is the real warlike activity, everything else is only its
auxiliary; let us therefore take an attentive look at its nature.

Combat means fighting, and in this the destruction or conquest of the
enemy is the object, and the enemy, in the particular combat, is the
armed force which stands opposed to us.

This is the simple idea; we shall return to it, but before we can do
that we must insert a series of others.

If we suppose the State and its military force as a unit, then the most
natural idea is to imagine the War also as one great combat, and in the
simple relations of savage nations it is also not much otherwise. But
our Wars are made up of a number of great and small simultaneous or
consecutive combats, and this severance of the activity into so many
separate actions is owing to the great multiplicity of the relations out
of which War arises with us.

In point of fact, the ultimate object of our Wars, the political one, is
not always quite a simple one; and even were it so, still the action is
bound up with such a number of conditions and considerations to be taken
into account, that the object can no longer be attained by one single
great act but only through a number of greater or smaller acts which are
bound up into a whole; each of these separate acts is therefore a part
of a whole, and has consequently a special object by which it is bound
to this whole.

We have already said that every strategic act can be referred to the
idea of a combat, because it is an employment of the military force,
and at the root of that there always lies the idea of fighting. We may
therefore reduce every military activity in the province of Strategy
to the unit of single combats, and occupy ourselves with the object
of these only; we shall get acquainted with these special objects by
degrees as we come to speak of the causes which produce them; here we
content ourselves with saying that every combat, great or small, has its
own peculiar object in subordination to the main object. If this is
the case then, the destruction and conquest of the enemy is only to be
regarded as the means of gaining this object; as it unquestionably is.

But this result is true only in its form, and important only on account
of the connection which the ideas have between themselves, and we have
only sought it out to get rid of it at once.

What is overcoming the enemy? Invariably the destruction of his military
force, whether it be by death, or wounds, or any means; whether it be
completely or only to such a degree that he can no longer continue
the contest; therefore as long as we set aside all special objects of
combats, we may look upon the complete or partial destruction of the
enemy as the only object of all combats.

Now we maintain that in the majority of cases, and especially in great
battles, the special object by which the battle is individualised
and bound up with the great whole is only a weak modification of that
general object, or an ancillary object bound up with it, important

enough to individualise the battle, but always insignificant in
comparison with that general object; so that if that ancillary object
alone should be obtained, only an unimportant part of the purpose of the
combat is fulfilled. If this assertion is correct, then we see that the
idea, according to which the destruction of the enemy's force is only
the means, and something else always the object, can only be true
in form, but, that it would lead to false conclusions if we did not
recollect that this destruction of the enemy's force is comprised in
that object, and that this object is only a weak modification of it.
Forgetfulness of this led to completely false views before the wars of
the last period, and created tendencies as well as fragments of
systems, in which theory thought it raised itself so much the more above
handicraft, the less it supposed itself to stand in need of the use of
the real instrument, that is the destruction of the enemy's force.

Certainly such a system could not have arisen unless supported by other
false suppositions, and unless in place of the destruction of the enemy,
other things had been substituted to which an efficacy was ascribed
which did not rightly belong to them. We shall attack these falsehoods
whenever occasion requires, but we could not treat of the combat without
claiming for it the real importance and value which belong to it, and
giving warning against the errors to which merely formal truth might
lead.

But now how shall we manage to show that in most cases, and in those of
most importance, the destruction of the enemy's Army is the chief thing?
How shall we manage to combat that extremely subtle idea, which supposes
it possible, through the use of a special artificial form, to effect
by a small direct destruction of the enemy's forces a much greater
destruction indirectly, or by means of small but extremely well-directed
blows to produce such paralysation of the enemy's forces, such a command
over the enemy's will, that this mode of proceeding is to be viewed as a
great shortening of the road? Undoubtedly a victory at one point may
be of more value than at another. Undoubtedly there is a scientific
arrangement of battles amongst themselves, even in Strategy, which is in
fact nothing but the Art of thus arranging them. To deny that is not
our intention, but we assert that the direct destruction of the enemy's
forces is everywhere predominant; we contend here for the overruling
importance of this destructive principle and nothing else.

We must, however, call to mind that we are now engaged with Strategy,
not with tactics, therefore we do not speak of the means which the
former may have of destroying at a small expense a large body of the
enemy's forces, but under direct destruction we understand the tactical
results, and that, therefore, our assertion is that only great tactical
results can lead to great strategical ones, or, as we have already
once before more distinctly expressed it, THE TACTICAL SUCCESSES are of
paramount importance in the conduct of War.

The proof of this assertion seems to us simple enough, it lies in the
time which every complicated (artificial) combination requires. The
question whether a simple attack, or one more carefully prepared,
i.e., more artificial, will produce greater effects, may undoubtedly
be decided in favour of the latter as long as the enemy is assumed to
remain quite passive. But every carefully combined attack requires time
for its preparation, and if a counter-stroke by the enemy intervenes,
our whole design may be upset. Now if the enemy should decide upon some
simple attack, which can be executed in a shorter time, then he gains
the initiative, and destroys the effect of the great plan. Therefore,
together with the expediency of a complicated attack we must consider
all the dangers which we run during its preparation, and should only
adopt it if there is no reason to fear that the enemy will disconcert
our scheme. Whenever this is the case we must ourselves choose the
simpler, i.e., quicker way, and lower our views in this sense as far as
the character, the relations of the enemy, and other circumstances may
render necessary. If we quit the weak impressions of abstract ideas and
descend to the region of practical life, then it is evident that a bold,
courageous, resolute enemy will not let us have time for wide-reaching

skilful combinations, and it is just against such a one we should require skill the most. By this it appears to us that the advantage of simple and direct results over those that are complicated is conclusively shown.

Our opinion is not on that account that the simple blow is the best, but that we must not lift the arm too far for the time given to strike, and that this condition will always lead more to direct conflict the more warlike our opponent is. Therefore, far from making it our aim to gain upon the enemy by complicated plans, we must rather seek to be beforehand with him by greater simplicity in our designs.

If we seek for the lowest foundation-stones of these converse propositions we find that in the one it is ability, in the other, courage. Now, there is something very attractive in the notion that a moderate degree of courage joined to great ability will produce greater effects than moderate ability with great courage. But unless we suppose these elements in a disproportionate relation, not logical, we have no right to assign to ability this advantage over courage in a field which is called danger, and which must be regarded as the true domain of courage.

After this abstract view we shall only add that experience, very far from leading to a different conclusion, is rather the sole cause which has impelled us in this direction, and given rise to such reflections.

Whoever reads history with a mind free from prejudice cannot fail to arrive at a conviction that of all military virtues, energy in the conduct of operations has always contributed the most to the glory and success of arms.

How we make good our principle of regarding the destruction of the enemy's force as the principal object, not only in the War as a whole but also in each separate combat, and how that principle suits all the forms and conditions necessarily demanded by the relations out of which War springs, the sequel will show. For the present all that we desire is to uphold its general importance, and with this result we return again to the combat.

CHAPTER IV. THE COMBAT IN GENERAL (CONTINUATION)

IN the last chapter we showed the destruction of the enemy as the true object of the combat, and we have sought to prove by a special consideration of the point, that this is true in the majority of cases, and in respect to the most important battles, because the destruction of the enemy's Army is always the preponderating object in War. The other objects which may be mixed up with this destruction of the enemy's force, and may have more or less influence, we shall describe generally in the next chapter, and become better acquainted with by degrees afterwards; here we divest the combat of them entirely, and look upon the destruction of the enemy as the complete and sufficient object of any combat.

What are we now to understand by destruction of the enemy's Army? A diminution of it relatively greater than that on our own side. If we have a great superiority in numbers over the enemy, then naturally the same absolute amount of loss on both sides is for us a smaller one than for him, and consequently may be regarded in itself as an advantage. As we are here considering the combat as divested of all (other) objects, we must also exclude from our consideration the case in which the combat is used only indirectly for a greater destruction of the enemy's force; consequently also, only that direct gain which has been made in the mutual process of destruction, is to be regarded as the object, for this is an absolute gain, which runs through the whole campaign, and at the end of it will always appear as pure profit. But every other kind of victory over our opponent will either have its motive in other objects,

which we have completely excluded here, or it will only yield a
temporary relative advantage. An example will make this plain.

If by a skilful disposition we have reduced our opponent to such a
dilemma, that he cannot continue the combat without danger, and after
some resistance he retires, then we may say, that we have conquered
him at that point; but if in this victory we have expended just as many
forces as the enemy, then in closing the account of the campaign, there
is no gain remaining from this victory, if such a result can be called
a victory. Therefore the overcoming the enemy, that is, placing him in
such a position that he must give up the fight, counts for nothing in
itself, and for that reason cannot come under the definition of object.
There remains, therefore, as we have said, nothing over except the
direct gain which we have made in the process of destruction; but to
this belong not only the losses which have taken place in the course of
the combat, but also those which, after the withdrawal of the conquered
part, take place as direct consequences of the same.

Now it is known by experience, that the losses in physical forces in the
course of a battle seldom present a great difference between victor and
vanquished respectively, often none at all, sometimes even one bearing
an inverse relation to the result, and that the most decisive losses
on the side of the vanquished only commence with the retreat, that is,
those which the conqueror does not share with him. The weak remains of
battalions already in disorder are cut down by cavalry, exhausted men
strew the ground, disabled guns and broken caissons are abandoned,
others in the bad state of the roads cannot be removed quickly enough,
and are captured by the enemy's troops, during the night numbers lose
their way, and fall defenceless into the enemy's hands, and thus the
victory mostly gains bodily substance after it is already decided. Here
would be a paradox, if it did not solve itself in the following manner.

The loss in physical force is not the only one which the two sides
suffer in the course of the combat; the moral forces also are shaken,
broken, and go to ruin. It is not only the loss in men, horses and guns,
but in order, courage, confidence, cohesion and plan, which come into
consideration when it is a question whether the fight can be still
continued or not. It is principally the moral forces which decide here,
and in all cases in which the conqueror has lost as heavily as the
conquered, it is these alone.

The comparative relation of the physical losses is difficult to estimate
in a battle, but not so the relation of the moral ones. Two things
principally make it known. The one is the loss of the ground on which
the fight has taken place, the other the superiority of the enemy's. The
more our reserves have diminished as compared with those of the enemy,
the more force we have used to maintain the equilibrium; in this at
once, an evident proof of the moral superiority of the enemy is given
which seldom fails to stir up in the soul of the Commander a certain
bitterness of feeling, and a sort of contempt for his own troops.
But the principal thing is, that men who have been engaged for a long
continuance of time are more or less like burnt-out cinders; their
ammunition is consumed; they have melted away to a certain extent;
physical and moral energies are exhausted, perhaps their courage is
broken as well. Such a force, irrespective of the diminution in its
number, if viewed as an organic whole, is very different from what it
was before the combat; and thus it is that the loss of moral force
may be measured by the reserves that have been used as if it were on a
foot-rule.

Lost ground and want of fresh reserves, are, therefore, usually the
principal causes which determine a retreat; but at the same time we by
no means exclude or desire to throw in the shade other reasons, which
may lie in the interdependence of parts of the Army, in the general
plan, &c.

Every combat is therefore the bloody and destructive measuring of the
strength of forces, physical and moral; whoever at the close has the

greatest amount of both left is the conqueror.

In the combat the loss of moral force is the chief cause of the
decision; after that is given, this loss continues to increase until it
reaches its culminating-point at the close of the whole act. This then
is the opportunity the victor should seize to reap his harvest by the
utmost possible restrictions of his enemy's forces, the real object of
engaging in the combat. On the beaten side, the loss of all order and
control often makes the prolongation of resistance by individual units,
by the further punishment they are certain to suffer, more injurious
than useful to the whole. The spirit of the mass is broken; the original
excitement about losing or winning, through which danger was forgotten,
is spent, and to the majority danger now appears no longer an appeal to
their courage, but rather the endurance of a cruel punishment. Thus the
instrument in the first moment of the enemy's victory is weakened and
blunted, and therefore no longer fit to repay danger by danger.

This period, however, passes; the moral forces of the conquered will
recover by degrees, order will be restored, courage will revive, and in
the majority of cases there remains only a small part of the superiority
obtained, often none at all. In some cases, even, although rarely, the
spirit of revenge and intensified hostility may bring about an opposite
result. On the other hand, whatever is gained in killed, wounded,
prisoners, and guns captured can never disappear from the account.

The losses in a battle consist more in killed and wounded; those
after the battle, more in artillery taken and prisoners. The first the
conqueror shares with the conquered, more or less, but the second not;
and for that reason they usually only take place on one side of the
conflict, at least, they are considerably in excess on one side.

Artillery and prisoners are therefore at all times regarded as the
true trophies of victory, as well as its measure, because through these
things its extent is declared beyond a doubt. Even the degree of moral
superiority may be better judged of by them than by any other relation,
especially if the number of killed and wounded is compared therewith;
and here arises a new power increasing the moral effects.

We have said that the moral forces, beaten to the ground in the
battle and in the immediately succeeding movements, recover themselves
gradually, and often bear no traces of injury; this is the case with
small divisions of the whole, less frequently with large divisions; it
may, however, also be the case with the main Army, but seldom or never
in the State or Government to which the Army belongs. These estimate the
situation more impartially, and from a more elevated point of view,
and recognise in the number of trophies taken by the enemy, and their
relation to the number of killed and wounded, only too easily and well,
the measure of their own weakness and inefficiency.

In point of fact, the lost balance of moral power must not be treated
lightly because it has no absolute value, and because it does not of
necessity appear in all cases in the amount of the results at the
final close; it may become of such excessive weight as to bring down
everything with an irresistible force. On that account it may often
become a great aim of the operations of which we shall speak elsewhere.
Here we have still to examine some of its fundamental relations.

The moral effect of a victory increases, not merely in proportion to
the extent of the forces engaged, but in a progressive ratio--that is
to say, not only in extent, but also in its intensity. In a beaten
detachment order is easily restored. As a single frozen limb is easily
revived by the rest of the body, so the courage of a defeated detachment
is easily raised again by the courage of the rest of the Army as soon
as it rejoins it. If, therefore, the effects of a small victory are not
completely done away with, still they are partly lost to the enemy. This
is not the case if the Army itself sustains a great defeat; then one
with the other fall together. A great fire attains quite a different
heat from several small ones.

Another relation which determines the moral value of a victory is the
numerical relation of the forces which have been in conflict with each
other. To beat many with few is not only a double success, but shows
also a greater, especially a more general superiority, which the
conquered must always be fearful of encountering again. At the same time
this influence is in reality hardly observable in such a case. In the
moment of real action, the notions of the actual strength of the
enemy are generally so uncertain, the estimate of our own commonly so
incorrect, that the party superior in numbers either does not admit the
disproportion, or is very far from admitting the full truth, owing to
which, he evades almost entirely the moral disadvantages which would
spring from it. It is only hereafter in history that the truth, long
suppressed through ignorance, vanity, or a wise discretion, makes its
appearance, and then it certainly casts a lustre on the Army and its
Leader, but it can then do nothing more by its moral influence for
events long past.

If prisoners and captured guns are those things by which the victory
principally gains substance, its true crystallisations, then the plan of
the battle should have those things specially in view; the destruction
of the enemy by death and wounds appears here merely as a means to an
end.

How far this may influence the dispositions in the battle is not an
affair of Strategy, but the decision to fight the battle is in intimate
connection with it, as is shown by the direction given to our forces,
and their general grouping, whether we threaten the enemy's flank or
rear, or he threatens ours. On this point, the number of prisoners and
captured guns depends very much, and it is a point which, in many cases,
tactics alone cannot satisfy, particularly if the strategic relations
are too much in opposition to it.

The risk of having to fight on two sides, and the still more dangerous
position of having no line of retreat left open, paralyse the movements
and the power of resistance; further, in case of defeat, they
increase the loss, often raising it to its extreme point, that is, to
destruction. Therefore, the rear being endangered makes defeat more
probable, and, at the same time, more decisive.

From this arises, in the whole conduct of the War, especially in great
and small combats, a perfect instinct to secure our own line of retreat
and to seize that of the enemy; this follows from the conception of
victory, which, as we have seen, is something beyond mere slaughter.

In this effort we see, therefore, the first immediate purpose in the
combat, and one which is quite universal. No combat is imaginable in
which this effort, either in its double or single form, does not go hand
in hand with the plain and simple stroke of force. Even the smallest
troop will not throw itself upon its enemy without thinking of its line
of retreat, and, in most cases, it will have an eye upon that of the
enemy also.

We should have to digress to show how often this instinct is prevented
from going the direct road, how often it must yield to the difficulties
arising from more important considerations: we shall, therefore, rest
contented with affirming it to be a general natural law of the combat.

It is, therefore, active; presses everywhere with its natural weight,
and so becomes the pivot on which almost all tactical and strategic
manoeuvres turn.

If we now take a look at the conception of victory as a whole, we find
in it three elements:--

1. The greater loss of the enemy in physical power.

2. In moral power.

3. His open avowal of this by the relinquishment of his intentions.

The returns made up on each side of losses in killed and wounded, are never exact, seldom truthful, and in most cases, full of intentional misrepresentations. Even the statement of the number of trophies is seldom to be quite depended on; consequently, when it is not considerable it may also cast a doubt even on the reality of the victory. Of the loss in moral forces there is no reliable measure, except in the trophies: therefore, in many cases, the giving up the contest is the only real evidence of the victory. It is, therefore, to be regarded as a confession of inferiority--as the lowering of the flag, by which, in this particular instance, right and superiority are conceded to the enemy, and this degree of humiliation and disgrace, which, however, must be distinguished from all the other moral consequences of the loss of equilibrium, is an essential part of the victory. It is this part alone which acts upon the public opinion outside the Army, upon the people and the Government in both belligerent States, and upon all others in any way concerned.

But renouncement of the general object is not quite identical with quitting the field of battle, even when the battle has been very obstinate and long kept up; no one says of advanced posts, when they retire after an obstinate combat, that they have given up their object; even in combats aimed at the destruction of the enemy's Army, the retreat from the battlefield is not always to be regarded as a relinquishment of this aim, as for instance, in retreats planned beforehand, in which the ground is disputed foot by foot; all this belongs to that part of our subject where we shall speak of the separate object of the combat; here we only wish to draw attention to the fact that in most cases the giving up of the object is very difficult to distinguish from the retirement from the battlefield, and that the impression produced by the latter, both in and out of the Army, is not to be treated lightly.

For Generals and Armies whose reputation is not made, this is in itself one of the difficulties in many operations, justified by circumstances when a succession of combats, each ending in retreat, may appear as a succession of defeats, without being so in reality, and when that appearance may exercise a very depressing influence. It is impossible for the retreating General by making known his real intentions to prevent the moral effect spreading to the public and his troops, for to do that with effect he must disclose his plans completely, which of course would run counter to his principal interests to too great a degree.

In order to draw attention to the special importance of this conception of victory we shall only refer to the battle of Soor,(*) the trophies from which were not important (a few thousand prisoners and twenty guns), and where Frederick proclaimed his victory by remaining for five days after on the field of battle, although his retreat into Silesia had been previously determined on, and was a measure natural to his whole situation. According to his own account, he thought he would hasten a peace by the moral effect of his victory. Now although a couple of other successes were likewise required, namely, the battle at Katholisch Hennersdorf, in Lusatia, and the battle of Kesseldorf, before this peace took place, still we cannot say that the moral effect of the battle of Soor was nil.

(*) Soor, or Sohr, Sept. 30, 1745; Hennersdorf, Nov. 23, 1745; Kealteldorf, Dec. 15, 1745, all in the Second Silesian War.

If it is chiefly the moral force which is shaken by defeat, and if the number of trophies reaped by the enemy mounts up to an unusual height, then the lost combat becomes a rout, but this is not the necessary consequence of every victory. A rout only sets in when the moral force of the defeated is very severely shaken then there often ensues a complete incapability of further resistance, and the whole action consists of giving way, that is of flight.

Jena and Belle Alliance were routs, but not so Borodino.

Although without pedantry we can here give no single line of separation, because the difference between the things is one of degrees, yet still the retention of the conception is essential as a central point to give clearness to our theoretical ideas and it is a want in our terminology that for a victory over the enemy tantamount to a rout, and a conquest of the enemy only tantamount to a simple victory, there is only one and the same word to use.

CHAPTER V. ON THE SIGNIFICATION OF THE COMBAT

HAVING in the preceding chapter examined the combat in its absolute form, as the miniature picture of the whole War, we now turn to the relations which it bears to the other parts of the great whole. First we inquire what is more precisely the signification of a combat.

As War is nothing else but a mutual process of destruction, then the most natural answer in conception, and perhaps also in reality, appears to be that all the powers of each party unite in one great volume and all results in one great shock of these masses. There is certainly much truth in this idea, and it seems to be very advisable that we should adhere to it and should on that account look upon small combats at first only as necessary loss, like the shavings from a carpenter's plane. Still, however, the thing cannot be settled so easily.

That a multiplication of combats should arise from a fractioning of forces is a matter of course, and the more immediate objects of separate combats will therefore come before us in the subject of a fractioning of forces; but these objects, and together with them, the whole mass of combats may in a general way be brought under certain classes, and the knowledge of these classes will contribute to make our observations more intelligible.

Destruction of the enemy's military forces is in reality the object of all combats; but other objects may be joined thereto, and these other objects may be at the same time predominant; we must therefore draw a distinction between those in which the destruction of the enemy's forces is the principal object, and those in which it is more the means. The destruction of the enemy's force, the possession of a place or the possession of some object may be the general motive for a combat, and it may be either one of these alone or several together, in which case however usually one is the principal motive. Now the two principal forms of War, the offensive and defensive, of which we shall shortly speak, do not modify the first of these motives, but they certainly do modify the other two, and therefore if we arrange them in a scheme they would appear thus:--

OFFENSIVE.	DEFENSIVE.
1. Destruction of enemy's force	1. Destruction of enemy's force.
2. Conquest of a place.	2. Defence of a place.
3. Conquest of some object.	3. Defence of some object.

These motives, however, do not seem to embrace completely the whole of the subject, if we recollect that there are reconnaissances and demonstrations, in which plainly none of these three points is the object of the combat. In reality we must, therefore, on this account be allowed a fourth class. Strictly speaking, in reconnaissances in which we wish the enemy to show himself, in alarms by which we wish to wear him out, in demonstrations by which we wish to prevent his leaving some point or to draw him off to another, the objects are all such as can only be attained indirectly and UNDER THE PRETEXT OF ONE OF THE THREE OBJECTS SPECIFIED IN THE TABLE, usually of the second; for the enemy

whose aim is to reconnoitre must draw up his force as if he really
intended to attack and defeat us, or drive us off, &c. &c. But this
pretended object is not the real one, and our present question is only
as to the latter; therefore, we must to the above three objects of the
offensive further add a fourth, which is to lead the enemy to make a
false conclusion. That offensive means are conceivable in connection
with this object, lies in the nature of the thing.

On the other hand we must observe that the defence of a place may be of
two kinds, either absolute, if as a general question the point is not to
be given up, or relative if it is only required for a certain time. The
latter happens perpetually in the combats of advanced posts and rear
guards.

That the nature of these different intentions of a combat must have an
essential influence on the dispositions which are its preliminaries, is
a thing clear in itself. We act differently if our object is merely to
drive an enemy's post out of its place from what we should if our object
was to beat him completely; differently, if we mean to defend a place
to the last extremity from what we should do if our design is only
to detain the enemy for a certain time. In the first case we trouble
ourselves little about the line of retreat, in the latter it is the
principal point, &c.

But these reflections belong properly to tactics, and are only
introduced here by way of example for the sake of greater clearness.
What Strategy has to say on the different objects of the combat will
appear in the chapters which touch upon these objects. Here we have only
a few general observations to make, first, that the importance of the
object decreases nearly in the order as they stand above, therefore,
that the first of these objects must always predominate in the great
battle; lastly, that the two last in a defensive battle are in reality
such as yield no fruit, they are, that is to say, purely negative,
and can, therefore, only be serviceable, indirectly, by facilitating
something else which is positive. IT IS, THEREFORE, A BAD SIGN OF THE
STRATEGIC SITUATION IF BATTLES OF THIS KIND BECOME TOO FREQUENT.

CHAPTER VI. DURATION OF THE COMBAT

IF we consider the combat no longer in itself but in relation to the
other forces of War, then its duration acquires a special importance.

This duration is to be regarded to a certain extent as a second
subordinate success. For the conqueror the combat can never be finished
too quickly, for the vanquished it can never last too long. A speedy
victory indicates a higher power of victory, a tardy decision is, on the
side of the defeated, some compensation for the loss.

This is in general true, but it acquires a practical importance in its
application to those combats, the object of which is a relative defence.

Here the whole success often lies in the mere duration. This is the
reason why we have included it amongst the strategic elements.

The duration of a combat is necessarily bound up with its essential
relations. These relations are, absolute magnitude of force, relation
of force and of the different arms mutually, and nature of the country.
Twenty thousand men do not wear themselves out upon one another as
quickly as two thousand: we cannot resist an enemy double or three times
our strength as long as one of the same strength; a cavalry combat is
decided sooner than an infantry combat; and a combat between infantry
only, quicker than if there is artillery(*) as well; in hills and
forests we cannot advance as quickly as on a level country; all this is
clear enough.

(*) The increase in the relative range of artillery and the

On War.txt
introduction of shrapnel has altogether modified this
conclusion.

From this it follows, therefore, that strength, relation of the three
arms, and position, must be considered if the combat is to fulfil an
object by its duration; but to set up this rule was of less importance
to us in our present considerations than to connect with it at once the
chief results which experience gives us on the subject.

Even the resistance of an ordinary Division of 8000 to 10,000 men of
all arms even opposed to an enemy considerably superior in numbers,
will last several hours, if the advantages of country are not too
preponderating, and if the enemy is only a little, or not at all,
superior in numbers, the combat will last half a day. A Corps of three
or four Divisions will prolong it to double the time; an Army of 80,000
or 100,000 to three or four times. Therefore the masses may be left to
themselves for that length of time, and no separate combat takes place
if within that time other forces can be brought up, whose co-operation
mingles then at once into one stream with the results of the combat
which has taken place.

These calculations are the result of experience; but it is important to
us at the same time to characterise more particularly the moment of the
decision, and consequently the termination.

CHAPTER VII. DECISION OF THE COMBAT

No battle is decided in a single moment, although in every battle there
arise moments of crisis, on which the result depends. The loss of a
battle is, therefore, a gradual falling of the scale. But there is in
every combat a point of time (*)

(*) Under the then existing conditions of armament
understood. This point is of supreme importance, as
practically the whole conduct of a great battle depends on a
correct solution of this question--viz., How long can a
given command prolong its resistance? If this is incorrectly
answered in practice--the whole manoeuvre depending on it
may collapse--e.g., Kouroupatkin at Liao-Yang, September
1904.

when it may be regarded as decided, in such a way that the renewal of
the fight would be a new battle, not a continuation of the old one. To
have a clear notion on this point of time, is very important, in
order to be able to decide whether, with the prompt assistance of
reinforcements, the combat can again be resumed with advantage.

Often in combats which are beyond restoration new forces are sacrificed
in vain; often through neglect the decision has not been seized when it
might easily have been secured. Here are two examples, which could not
be more to the point:

When the Prince of Hohenlohe, in 1806, at Jena,(*) with 35,000 men
opposed to from 60,000 to 70,000 under Buonaparte, had accepted battle,
and lost it--but lost it in such a way that the 35,000 might be regarded
as dissolved--General Ruchel undertook to renew the fight with about
12,000; the consequence was that in a moment his force was scattered in
like manner.

(*) October 14, 1806.

On the other hand, on the same day at Auerstadt, the Prussians
maintained a combat with 25,000, against Davoust, who had 28,000, until
mid-day, without success, it is true, but still without the force being
reduced to a state of dissolution without even greater loss than the
enemy, who was very deficient in cavalry;--but they neglected to use the
Page 134

reserve of 18,000, under General Kalkreuth, to restore the battle which,
under these circumstances, it would have been impossible to lose.

Each combat is a whole in which the partial combats combine themselves
into one total result. In this total result lies the decision of the
combat. This success need not be exactly a victory such as we have
denoted in the sixth chapter, for often the preparations for that have
not been made, often there is no opportunity if the enemy gives way too
soon, and in most cases the decision, even when the resistance has been
obstinate, takes place before such a degree of success is attained as
would completely satisfy the idea of a victory.

We therefore ask, which is commonly the moment of the decision, that
is to say, that moment when a fresh, effective, of course not
disproportionate, force, can no longer turn a disadvantageous battle?

If we pass over false attacks, which in accordance with their nature are
properly without decision, then,

1. If the possession of a movable object was the object of the combat,
the loss of the same is always the decision.

2. If the possession of ground was the object of the combat, then the
decision generally lies in its loss. Still not always, only if this
ground is of peculiar strength, ground which is easy to pass over,
however important it may be in other respects, can be re-taken without
much danger.

3. But in all other cases, when these two circumstances have not already
decided the combat, therefore, particularly in case the destruction of
the enemy's force is the principal object, the decision is reached at
that moment when the conqueror ceases to feel himself in a state of
disintegration, that is, of unserviceableness to a certain extent, when
therefore, there is no further advantage in using the successive efforts
spoken of in the twelfth chapter of the third book. On this ground we
have given the strategic unity of the battle its place here.

A battle, therefore, in which the assailant has not lost his condition
of order and perfect efficiency at all, or, at least, only in a small
part of his force, whilst the opposing forces are, more or less,
disorganised throughout, is also not to be retrieved; and just as little
if the enemy has recovered his efficiency.

The smaller, therefore, that part of a force is which has really been
engaged, the greater that portion which as reserve has contributed to
the result only by its presence. So much the less will any new force of
the enemy wrest again the victory from our hands, and that Commander who
carries out to the furthest with his Army the principle of conducting
the combat with the greatest economy of forces, and making the most of
the moral effect of strong reserves, goes the surest way to victory.
We must allow that the French, in modern times, especially when led by
Buonaparte, have shown a thorough mastery in this.

Further, the moment when the crisis-stage of the combat ceases with
the conqueror, and his original state of order is restored, takes place
sooner the smaller the unit he controls. A picket of cavalry pursuing an
enemy at full gallop will in a few minutes resume its proper order, and
the crisis ceases. A whole regiment of cavalry requires a longer time.
It lasts still longer with infantry, if extended in single lines of
skirmishers, and longer again with Divisions of all arms, when it
happens by chance that one part has taken one direction and another part
another direction, and the combat has therefore caused a loss of the
order of formation, which usually becomes still worse from no part
knowing exactly where the other is. Thus, therefore, the point of time
when the conqueror has collected the instruments he has been using, and
which are mixed up and partly out of order, the moment when he has in
some measure rearranged them and put them in their proper places, and
thus brought the battle-workshop into a little order, this moment, we

On War.txt
say, is always later, the greater the total force.

Again, this moment comes later if night overtakes the conqueror in the
crisis, and, lastly, it comes later still if the country is broken and
thickly wooded. But with regard to these two points, we must observe
that night is also a great means of protection, and it is only seldom
that circumstances favour the expectation of a successful result from
a night attack, as on March 10, 1814, at Laon,(*) where York against
Marmont gives us an example completely in place here. In the same way a
wooded and broken country will afford protection against a reaction to
those who are engaged in the long crisis of victory. Both, therefore,
the night as well as the wooded and broken country are obstacles
which make the renewal of the same battle more difficult instead of
facilitating it.

(*) The celebrated charge at night upon Marmont's Corps.

Hitherto, we have considered assistance arriving for the losing side
as a mere increase of force, therefore, as a reinforcement coming up
directly from the rear, which is the most usual case. But the case is
quite different if these fresh forces come upon the enemy in flank or
rear.

On the effect of flank or rear attacks so far as they belong to
Strategy, we shall speak in another place: such a one as we have here
in view, intended for the restoration of the combat, belongs chiefly to
tactics, and is only mentioned because we are here speaking of tactical
results, our ideas, therefore, must trench upon the province of tactics.

By directing a force against the enemy's flank and rear its efficacy may
be much intensified; but this is so far from being a necessary result
always that the efficacy may, on the other hand, be just as much
weakened. The circumstances under which the combat has taken place
decide upon this part of the plan as well as upon every other, without
our being able to enter thereupon here. But, at the same time, there are
in it two things of importance for our subject: first, FLANK AND REAR
ATTACKS HAVE, AS A RULE, A MORE FAVOURABLE EFFECT ON THE CONSEQUENCES
OF THE DECISION THAN UPON THE DECISION ITSELF. Now as concerns the
retrieving a battle, the first thing to be arrived at above all is a
favourable decision and not magnitude of success. In this view one would
therefore think that a force which comes to re-establish our combat
is of less assistance if it falls upon the enemy in flank and rear,
therefore separated from us, than if it joins itself to us directly;
certainly, cases are not wanting where it is so, but we must say that
the majority are on the other side, and they are so on account of the
second point which is here important to us.

This second point IS THE MORAL EFFECT OF THE SURPRISE, WHICH, AS A RULE,
A REINFORCEMENT COMING UP TO RE-ESTABLISH A COMBAT HAS GENERALLY IN ITS
FAVOUR. Now the effect of a surprise is always heightened if it takes
place in the flank or rear, and an enemy completely engaged in the
crisis of victory in his extended and scattered order, is less in a
state to counteract it. who does not feel that an attack in flank or
rear, which at the commencement of the battle, when the forces
are concentrated and prepared for such an event would be of little
importance, gains quite another weight in the last moment of the combat.

We must, therefore, at once admit that in most cases a reinforcement
coming up on the flank or rear of the enemy will be more efficacious,
will be like the same weight at the end of a longer lever, and therefore
that under these circumstances, we may undertake to restore the battle
with the same force which employed in a direct attack would be quite
insufficient. Here results almost defy calculation, because the moral
forces gain completely the ascendency. This is therefore the right field
for boldness and daring.

The eye must, therefore, be directed on all these objects, all these
moments of co-operating forces must be taken into consideration, when we

have to decide in doubtful cases whether or not it is still possible to
restore a combat which has taken an unfavourable turn.

If the combat is to be regarded as not yet ended, then the new contest
which is opened by the arrival of assistance fuses into the former;
therefore they flow together into one common result, and the first
disadvantage vanishes completely out of the calculation. But this is not
the case if the combat was already decided; then there are two results
separate from each other. Now if the assistance which arrives is only of
a relative strength, that is, if it is not in itself alone a match for
the enemy, then a favourable result is hardly to be expected from this
second combat: but if it is so strong that it can undertake the second
combat without regard to the first, then it may be able by a favourable
issue to compensate or even overbalance the first combat, but never to
make it disappear altogether from the account.

At the battle of Kunersdorf,(*) Frederick the Great at the first onset
carried the left of the Russian position, and took seventy pieces of
artillery; at the end of the battle both were lost again, and the whole
result of the first combat was wiped out of the account. Had it been
possible to stop at the first success, and to put off the second part
of the battle to the coming day, then, even if the King had lost it, the
advantages of the first would always have been a set off to the second.

        (*) August 12, 1759.

But when a battle proceeding disadvantageously is arrested and turned
before its conclusion, its minus result on our side not only disappears
from the account, but also becomes the foundation of a greater victory.
If, for instance, we picture to ourselves exactly the tactical course
of the battle, we may easily see that until it is finally concluded all
successes in partial combats are only decisions in suspense, which by
the capital decision may not only be destroyed, but changed into the
opposite. The more our forces have suffered, the more the enemy will
have expended on his side; the greater, therefore, will be the crisis
for the enemy, and the more the superiority of our fresh troops will
tell. If now the total result turns in our favour, if we wrest from the
enemy the field of battle and recover all the trophies again, then all
the forces which he has sacrificed in obtaining them become sheer gain
for us, and our former defeat becomes a stepping-stone to a greater
triumph. The most brilliant feats which with victory the enemy would
have so highly prized that the loss of forces which they cost would have
been disregarded, leave nothing now behind but regret at the sacrifice
entailed. Such is the alteration which the magic of victory and the
curse of defeat produces in the specific weight of the same elements.

Therefore, even if we are decidedly superior in strength, and are able
to repay the enemy his victory by a greater still, it is always better
to forestall the conclusion of a disadvantageous combat, if it is
of proportionate importance, so as to turn its course rather than to
deliver a second battle.

Field-Marshal Daun attempted in the year 1760 to come to the assistance
of General Laudon at Leignitz, whilst the battle lasted; but when he
failed, he did not attack the King next day, although he did not want
for means to do so.

For these reasons serious combats of advance guards which precede a
battle are to be looked upon only as necessary evils, and when not
necessary they are to be avoided.(*)

        (*) This, however, was not Napoleon's view. A vigorous
        attack of his advance guard he held to be necessary always,
        to fix the enemy's attention and "paralyse his independent
        will-power." It was the failure to make this point which, in
        August 1870, led von Moltke repeatedly into the very jaws of
        defeat, from which only the lethargy of Bazaine on the one
        hand and the initiative of his subordinates, notably of von

Alvensleben, rescued him. This is the essence of the new Strategic Doctrine of the French General Staff. See the works of Bonnal, Foch, &C.--EDITOR

We have still another conclusion to examine.

If on a regular pitched battle, the decision has gone against one, this does not constitute a motive for determining on a new one. The determination for this new one must proceed from other relations. This conclusion, however, is opposed by a moral force, which we must take into account: it is the feeling of rage and revenge. From the oldest Field-Marshal to the youngest drummer-boy this feeling is general, and, therefore, troops are never in better spirits for fighting than when they have to wipe out a stain. This is, however, only on the supposition that the beaten portion is not too great in proportion to the whole, because otherwise the above feeling is lost in that of powerlessness.

There is therefore a very natural tendency to use this moral force to repair the disaster on the spot, and on that account chiefly to seek another battle if other circumstances permit. It then lies in the nature of the case that this second battle must be an offensive one.

In the catalogue of battles of second-rate importance there are many examples to be found of such retaliatory battles; but great battles have generally too many other determining causes to be brought on by this weaker motive.

Such a feeling must undoubtedly have led the noble Bluecher with his third Corps to the field of battle on February 14, 1814, when the other two had been beaten three days before at Montmirail. Had he known that he would have come upon Buonaparte in person, then, naturally, preponderating reasons would have determined him to put off his revenge to another day: but he hoped to revenge himself on Marmont, and instead of gaining the reward of his desire for honourable satisfaction, he suffered the penalty of his erroneous calculation.

On the duration of the combat and the moment of its decision depend the distances from each other at which those masses should be placed which are intended to fight IN CONJUNCTION WITH each other. This disposition would be a tactical arrangement in so far as it relates to one and the same battle; it can, however, only be regarded as such, provided the position of the troops is so compact that two separate combats cannot be imagined, and consequently that the space which the whole occupies can be regarded strategically as a mere point. But in War, cases frequently occur where even those forces intended to fight IN UNISON must be so far separated from each other that while their union for one common combat certainly remains the principal object, still the occurrence of separate combats remains possible. Such a disposition is therefore strategic.

Dispositions of this kind are: marches in separate masses and columns, the formation of advance guards, and flanking columns, also the grouping of reserves intended to serve as supports for more than one strategic point; the concentration of several Corps from widely extended cantonments, &c. &c. We can see that the necessity for these arrangements may constantly arise, and may consider them something like the small change in the strategic economy, whilst the capital battles, and all that rank with them are the gold and silver pieces.

CHAPTER VIII. MUTUAL UNDERSTANDING AS TO A BATTLE

NO battle can take place unless by mutual consent; and in this idea, which constitutes the whole basis of a duel, is the root of a certain phraseology used by historical writers, which leads to many indefinite and false conceptions.

According to the view of the writers to whom we refer, it has frequently

happened that one Commander has offered battle to the other, and the
latter has not accepted it.

But the battle is a very modified duel, and its foundation is not merely
in the mutual wish to fight, that is in consent, but in the objects
which are bound up with the battle: these belong always to a greater
whole, and that so much the more, as even the whole war considered as
a "combat-unit" has political objects and conditions which belong to a
higher standpoint. The mere desire to conquer each other therefore falls
into quite a subordinate relation, or rather it ceases completely to be
anything of itself, and only becomes the nerve which conveys the impulse
of action from the higher will.

Amongst the ancients, and then again during the early period of standing
Armies, the expression that we had offered battle to the enemy in vain,
had more sense in it than it has now. By the ancients everything was
constituted with a view to measuring each other's strength in the open
field free from anything in the nature of a hindrance,(*) and the whole
Art of War consisted in the organisation, and formation of the Army,
that is in the order of battle.

> (*) Note the custom of sending formal challenges, fix time
> and place for action, and "enhazelug" the battlefield in
> Anglo-Saxon times.--ED.

Now as their Armies regularly entrenched themselves in their camps,
therefore the position in a camp was regarded as something unassailable,
and a battle did not become possible until the enemy left his camp, and
placed himself in a practicable country, as it were entered the lists.

If therefore we hear about Hannibal having offered battle to Fabius
in vain, that tells us nothing more as regards the latter than that
a battle was not part of his plan, and in itself neither proves the
physical nor moral superiority of Hannibal; but with respect to him the
expression is still correct enough in the sense that Hannibal really
wished a battle.

In the early period of modern Armies, the relations were similar in
great combats and battles. That is to say, great masses were brought
into action, and managed throughout it by means of an order of battle,
which like a great helpless whole required a more or less level plain
and was neither suited to attack, nor yet to defence in a broken, close
or even mountainous country. The defender therefore had here also to
some extent the means of avoiding battle. These relations although
gradually becoming modified, continued until the first Silesian War, and
it was not until the Seven Years' War that attacks on an enemy posted in a
difficult country gradually became feasible, and of ordinary occurrence:
ground did not certainly cease to be a principle of strength to those
making use of its aid, but it was no longer a charmed circle, which shut
out the natural forces of War.

During the past thirty years War has perfected itself much more in this
respect, and there is no longer anything which stands in the way of a
General who is in earnest about a decision by means of battle; he can
seek out his enemy, and attack him: if he does not do so he cannot
take credit for having wished to fight, and the expression he offered
a battle which his opponent did not accept, therefore now means nothing
more than that he did not find circumstances advantageous enough for a
battle, an admission which the above expression does not suit, but which
it only strives to throw a veil over.

It is true the defensive side can no longer refuse a battle, yet he may
still avoid it by giving up his position, and the role with which that
position was connected: this is however half a victory for the offensive
side, and an acknowledgment of his superiority for the present.

This idea in connection with the cartel of defiance can therefore no
longer be made use of in order by such rhodomontade to qualify the

inaction of him whose part it is to advance, that is, the offensive. The defender who as long as he does not give way, must have the credit of willing the battle, may certainly say, he has offered it if he is not attacked, if that is not understood of itself.

But on the other hand, he who now wishes to, and can retreat cannot easily be forced to give battle. Now as the advantages to the aggressor from this retreat are often not sufficient, and a substantial victory is a matter of urgent necessity for him, in that way the few means which there are to compel such an opponent also to give battle are often sought for and applied with particular skill.

The principal means for this are--first SURROUNDING the enemy so as to make his retreat impossible, or at least so difficult that it is better for him to accept battle; and, secondly, SURPRISING him. This last way, for which there was a motive formerly in the extreme difficulty of all movements, has become in modern times very inefficacious.

From the pliability and manoeuvring capabilities of troops in the present day, one does not hesitate to commence a retreat even in sight of the enemy, and only some special obstacles in the nature of the country can cause serious difficulties in the operation.

As an example of this kind the battle of Neresheim may be given, fought by the Archduke Charles with Moreau in the Rauhe Alp, August 11, 1796, merely with a view to facilitate his retreat, although we freely confess we have never been able quite to understand the argument of the renowned general and author himself in this case.

The battle of Rosbach(*) is another example, if we suppose the commander of the allied army had not really the intention of attacking Frederick the Great.

(*) November 5, 1757.

Of the battle of Soor,(*) the King himself says that it was only fought because a retreat in the presence of the enemy appeared to him a critical operation; at the same time the King has also given other reasons for the battle.

(*) Or Sohr, September 30, 1745.

On the whole, regular night surprises excepted, such cases will always be of rare occurrence, and those in which an enemy is compelled to fight by being practically surrounded, will happen mostly to single corps only, like Mortier's at Durrenstein 1809, and Vandamme at Kulm, 1813.

CHAPTER IX. THE BATTLE(*)

(*) Clausewitz still uses the word "die Hauptschlacht" but modern usage employs only the word "die Schlacht" to designate the decisive act of a whole campaign--encounters arising from the collision or troops marching towards the strategic culmination of each portion or the campaign are spoken of either as "Treffen," i.e., "engagements" or "Gefecht," i.e., "combat" or "action." Thus technically, Gravelotte was a "Schlacht," i.e., "battle," but Spicheren, Woerth, Borny, even Vionville were only "Treffen."

ITS DECISION

WHAT is a battle? A conflict of the main body, but not an unimportant one about a secondary object, not a mere attempt which is given up when we see betimes that our object is hardly within our reach: it is a conflict waged with all our forces for the attainment of a decisive victory.

Minor objects may also be mixed up with the principal object, and it will take many different tones of colour from the circumstances out of which it originates, for a battle belongs also to a greater whole of which it is only a part, but because the essence of War is conflict, and the battle is the conflict of the main Armies, it is always to be regarded as the real centre of gravity of the War, and therefore its distinguishing character is, that unlike all other encounters, it is arranged for, and undertaken with the sole purpose of obtaining a decisive victory.

This has an influence on the MANNER OF ITS DECISION, on the EFFECT OF THE VICTORY CONTAINED IN IT, and determines THE VALUE WHICH THEORY IS TO ASSIGN TO IT AS A MEANS TO AN END.

On that account we make it the subject of our special consideration, and at this stage before we enter upon the special ends which may be bound up with it, but which do not essentially alter its character if it really deserves to be termed a battle.

If a battle takes place principally on its own account, the elements of its decision must be contained in itself; in other words, victory must be striven for as long as a possibility or hope remains. It must not, therefore, be given up on account of secondary circumstances, but only and alone in the event of the forces appearing completely insufficient.

Now how is that precise moment to be described?

If a certain artificial formation and cohesion of an Army is the principal condition under which the bravery of the troops can gain a victory, as was the case during a great part of the period of the modern Art of War, THEN THE BREAKING UP OF THIS FORMATION is the decision. A beaten wing which is put out of joint decides the fate of all that was connected with it. If as was the case at another time the essence of the defence consists in an intimate alliance of the Army with the ground on which it fights and its obstacles, so that Army and position are only one, then the CONQUEST of AN ESSENTIAL POINT in this position is the decision. It is said the key of the position is lost, it cannot therefore be defended any further; the battle cannot be continued. In both cases the beaten Armies are very much like the broken strings of an instrument which cannot do their work.

That geometrical as well as this geographical principle which had a tendency to place an Army in a state of crystallising tension which did not allow of the available powers being made use of up to the last man, have at least so far lost their influence that they no longer predominate. Armies are still led into battle in a certain order, but that order is no longer of decisive importance; obstacles of ground are also still turned to account to strengthen a position, but they are no longer the only support.

We attempted in the second chapter of this book to take a general view of the nature of the modern battle. According to our conception of it, the order of battle is only a disposition of the forces suitable to the convenient use of them, and the course of the battle a mutual slow wearing away of these forces upon one another, to see which will have soonest exhausted his adversary.

The resolution therefore to give up the fight arises, in a battle more than in any other combat, from the relation of the fresh reserves remaining available; for only these still retain all their moral vigour, and the cinders of the battered, knocked-about battalions, already burnt out in the destroying element, must not be placed on a level with them; also lost ground as we have elsewhere said, is a standard of lost moral force; it therefore comes also into account, but more as a sign of loss suffered than for the loss itself, and the number of fresh reserves is always the chief point to be looked at by both Commanders.

On War.txt

In general, an action inclines in one direction from the very
commencement, but in a manner little observable. This direction is also
frequently given in a very decided manner by the arrangements which have
been made previously, and then it shows a want of discernment in that
General who commences battle under these unfavourable circumstances
without being aware of them. Even when this does not occur it lies in
the nature of things that the course of a battle resembles rather a slow
disturbance of equilibrium which commences soon, but as we have said
almost imperceptibly at first, and then with each moment of time becomes
stronger and more visible, than an oscillating to and fro, as those who
are misled by mendacious descriptions usually suppose.

But whether it happens that the balance is for a long time little
disturbed, or that even after it has been lost on one side it rights
itself again, and is then lost on the other side, it is certain at all
events that in most instances the defeated General foresees his fate
long before he retreats, and that cases in which some critical event
acts with unexpected force upon the course of the whole have their
existence mostly in the colouring with which every one depicts his lost
battle.

We can only here appeal to the decision of unprejudiced men of
experience, who will, we are sure, assent to what we have said, and
answer for us to such of our readers as do not know War from their own
experience. To develop the necessity of this course from the nature of
the thing would lead us too far into the province of tactics, to which
this branch of the subject belongs; we are here only concerned with its
results.

If we say that the defeated General foresees the unfavourable result
usually some time before he makes up his mind to give up the battle, we
admit that there are also instances to the contrary, because otherwise
we should maintain a proposition contradictory in itself. If at the
moment of each decisive tendency of a battle it should be considered as
lost, then also no further forces should be used to give it a turn, and
consequently this decisive tendency could not precede the retreat by
any length of time. Certainly there are instances of battles which after
having taken a decided turn to one side have still ended in favour
of the other; but they are rare, not usual; these exceptional cases,
however, are reckoned upon by every General against whom fortune
declares itself, and he must reckon upon them as long as there remains
a possibility of a turn of fortune. He hopes by stronger efforts, by
raising the remaining moral forces, by surpassing himself, or also by
some fortunate chance that the next moment will bring a change, and
pursues this as far as his courage and his judgment can agree. We shall
have something more to say on this subject, but before that we must show
what are the signs of the scales turning.

The result of the whole combat consists in the sum total of the results
of all partial combats; but these results of separate combats are
settled by different considerations.

First by the pure moral power in the mind of the leading officers. If a
General of Division has seen his battalions forced to succumb, it will
have an influence on his demeanour and his reports, and these again will
have an influence on the measures of the Commander-in-Chief; therefore
even those unsuccessful partial combats which to all appearance are
retrieved, are not lost in their results, and the impressions from them
sum themselves up in the mind of the Commander without much trouble, and
even against his will.

Secondly, by the quicker melting away of our troops, which can be easily
estimated in the slow and relatively(*) little tumultuary course of our
battles.

(*) Relatively, that is say to the shock of former days.

Thirdly, by lost ground.

All these things serve for the eye of the General as a compass to tell
the course of the battle in which he is embarked. If whole batteries
have been lost and none of the enemy's taken; if battalions have been
overthrown by the enemy's cavalry, whilst those of the enemy everywhere
present impenetrable masses; if the line of fire from his order of
battle wavers involuntarily from one point to another; if fruitless
efforts have been made to gain certain points, and the assaulting
battalions each, time been scattered by well-directed volleys of grape
and case;--if our artillery begins to reply feebly to that of the
enemy--if the battalions under fire diminish unusually, fast, because
with the wounded crowds of unwounded men go to the rear;--if single
Divisions have been cut off and made prisoners through the disruption of
the plan of the battle;--if the line of retreat begins to be endangered:
the Commander may tell very well in which direction he is going with
his battle. The longer this direction continues, the more decided it
becomes, so much the more difficult will be the turning, so much the
nearer the moment when he must give up the battle. We shall now make
some observations on this moment.

We have already said more than once that the final decision is ruled
mostly by the relative number of the fresh reserves remaining at the
last; that Commander who sees his adversary is decidedly superior to him
in this respect makes up his mind to retreat. It is the characteristic
of modern battles that all mischances and losses which take place in
the course of the same can be retrieved by fresh forces, because the
arrangement of the modern order of battle, and the way in which troops
are brought into action, allow of their use almost generally, and in
each position. So long, therefore, as that Commander against whom the
issue seems to declare itself still retains a superiority in reserve
force, he will not give up the day. But from the moment that his
reserves begin to become weaker than his enemy's, the decision may be
regarded as settled, and what he now does depends partly on special
circumstances, partly on the degree of courage and perseverance which he
personally possesses, and which may degenerate into foolish obstinacy.
How a Commander can attain to the power of estimating correctly the
still remaining reserves on both sides is an affair of skilful practical
genius, which does not in any way belong to this place; we keep
ourselves to the result as it forms itself in his mind. But this
conclusion is still not the moment of decision properly, for a motive
which only arises gradually does not answer to that, but is only a
general motive towards resolution, and the resolution itself requires
still some special immediate causes. Of these there are two chief ones
which constantly recur, that is, the danger of retreat, and the arrival
of night.

If the retreat with every new step which the battle takes in its course
becomes constantly in greater danger, and if the reserves are so much
diminished that they are no longer adequate to get breathing room, then
there is nothing left but to submit to fate, and by a well-conducted
retreat to save what, by a longer delay ending in flight and disaster,
would be lost.

But night as a rule puts an end to all battles, because a night combat
holds out no hope of advantage except under particular circumstances;
and as night is better suited for a retreat than the day, so, therefore,
the Commander who must look at the retreat as a thing inevitable, or as
most probable, will prefer to make use of the night for his purpose.

That there are, besides the above two usual and chief causes, yet many
others also, which are less or more individual and not to be overlooked,
is a matter of course; for the more a battle tends towards a complete
upset of equilibrium the more sensible is the influence of each partial
result in hastening the turn. Thus the loss of a battery, a successful
charge of a couple of regiments of cavalry, may call into life the
resolution to retreat already ripening.

As a conclusion to this subject, we must dwell for a moment on the point at which the courage of the Commander engages in a sort of conflict with his reason.

If, on the one hand the overbearing pride of a victorious conqueror, if the inflexible will of a naturally obstinate spirit, if the strenuous resistance of noble feelings will not yield the battlefield, where they must leave their honour, yet on the other hand, reason counsels not to give up everything, not to risk the last upon the game, but to retain as much over as is necessary for an orderly retreat. However highly we must esteem courage and firmness in war, and however little prospect there is of victory to him who cannot resolve to seek it by the exertion of all his power, still there is a point beyond which perseverance can only be termed desperate folly, and therefore can meet with no approbation from any critic. In the most celebrated of all battles, that of Belle-Alliance, Buonaparte used his last reserve in an effort to retrieve a battle which was past being retrieved. He spent his last farthing, and then, as a beggar, abandoned both the battle-field and his crown.

CHAPTER X. EFFECTS OF VICTORY (continuation)

ACCORDING to the point from which our view is taken, we may feel as much astonished at the extraordinary results of some great battles as at the want of results in others. We shall dwell for a moment on the nature of the effect of a great victory.

Three things may easily be distinguished here: the effect upon the instrument itself, that is, upon the Generals and their Armies; the effect upon the States interested in the War; and the particular result of these effects as manifested in the subsequent course of the campaign.

If we only think of the trifling difference which there usually is between victor and vanquished in killed, wounded, prisoners, and artillery lost on the field of battle itself, the consequences which are developed out of this insignificant point seem often quite incomprehensible, and yet, usually, everything only happens quite naturally.

We have already said in the seventh chapter that the magnitude of a victory increases not merely in the same measure as the vanquished forces increase in number, but in a higher ratio. The moral effects resulting from the issue of a great battle are greater on the side of the conquered than on that of the conqueror: they lead to greater losses in physical force, which then in turn react on the moral element, and so they go on mutually supporting and intensifying each other. On this moral effect we must therefore lay special weight. It takes an opposite direction on the one side from that on the other; as it undermines the energies of the conquered so it elevates the powers and energy of the conqueror. But its chief effect is upon the vanquished, because here it is the direct cause of fresh losses, and besides it is homogeneous in nature with danger, with the fatigues, the hardships, and generally with all those embarrassing circumstances by which War is surrounded, therefore enters into league with them and increases by their help, whilst with the conqueror all these things are like weights which give a higher swing to his courage. It is therefore found, that the vanquished sinks much further below the original line of equilibrium than the conqueror raises himself above it; on this account, if we speak of the effects of victory we allude more particularly to those which manifest themselves in the army. If this effect is more powerful in an important combat than in a smaller one, so again it is much more powerful in a great battle than in a minor one. The great battle takes place for the sake of itself, for the sake of the victory which it is to give, and which is sought for with the utmost effort. Here on this spot, in this very hour, to conquer the enemy is the purpose in which the plan of the War with all its threads converges, in which all distant hopes, all dim glimmerings of the future meet, fate steps in before us to give an

answer to the bold question.--This is the state of mental tension
not only of the Commander but of his whole Army down to the lowest
waggon-driver, no doubt in decreasing strength but also in decreasing
importance.

According to the nature of the thing, a great battle has never at any
time been an unprepared, unexpected, blind routine service, but a grand
act, which, partly of itself and partly from the aim of the Commander,
stands out from amongst the mass of ordinary efforts, sufficiently to
raise the tension of all minds to a higher degree. But the higher this
tension with respect to the issue, the more powerful must be the effect
of that issue.

Again, the moral effect of victory in our battles is greater than it was
in the earlier ones of modern military history. If the former are as we
have depicted them, a real struggle of forces to the utmost, then the
sum total of all these forces, of the physical as well as the moral,
must decide more than certain special dispositions or mere chance.

A single fault committed may be repaired next time; from good fortune
and chance we can hope for more favour on another occasion; but the sum
total of moral and physical powers cannot be so quickly altered, and,
therefore, what the award of a victory has decided appears of much
greater importance for all futurity. Very probably, of all concerned in
battles, whether in or out of the Army, very few have given a thought
to this difference, but the course of the battle itself impresses on the
minds of all present in it such a conviction, and the relation of this
course in public documents, however much it may be coloured by twisting
particular circumstances, shows also, more or less, to the world at
large that the causes were more of a general than of a particular
nature.

He who has not been present at the loss of a great battle will have
difficulty in forming for himself a living or quite true idea of it, and
the abstract notions of this or that small untoward affair will never
come up to the perfect conception of a lost battle. Let us stop a moment
at the picture.

The first thing which overpowers the imagination--and we may indeed say,
also the understanding--is the diminution of the masses; then the loss
of ground, which takes place always, more or less, and, therefore, on
the side of the assailant also, if he is not fortunate; then the rupture
of the original formation, the jumbling together of troops, the risks
of retreat, which, with few exceptions may always be seen sometimes in
a less sometimes in a greater degree; next the retreat, the most part of
which commences at night, or, at least, goes on throughout the night.
On this first march we must at once leave behind, a number of men
completely worn out and scattered about, often just the bravest, who
have been foremost in the fight who held out the longest: the feeling
of being conquered, which only seized the superior officers on the
battlefield, now spreads through all ranks, even down to the common
soldiers, aggravated by the horrible idea of being obliged to leave in
the enemy's hands so many brave comrades, who but a moment since were of
such value to us in the battle, and aggravated by a rising distrust
of the chief, to whom, more or less, every subordinate attributes as
a fault the fruitless efforts he has made; and this feeling of being
conquered is no ideal picture over which one might become master; it is
an evident truth that the enemy is superior to us; a truth of which
the causes might have been so latent before that they were not to be
discovered, but which, in the issue, comes out clear and palpable, or
which was also, perhaps, before suspected, but which in the want of
any certainty, we had to oppose by the hope of chance, reliance on
good fortune, Providence or a bold attitude. Now, all this has proved
insufficient, and the bitter truth meets us harsh and imperious.

All these feelings are widely different from a panic, which in an
army fortified by military virtue never, and in any other, only
exceptionally, follows the loss of a battle. They must arise even in

the best of Armies, and although long habituation to War and victory
together with great confidence in a Commander may modify them a little
here and there, they are never entirely wanting in the first moment.
They are not the pure consequences of lost trophies; these are usually
lost at a later period, and the loss of them does not become generally
known so quickly; they will therefore not fail to appear even when the
scale turns in the slowest and most gradual manner, and they constitute
that effect of a victory upon which we can always count in every case.

We have already said that the number of trophies intensifies this
effect.

It is evident that an Army in this condition, looked at as an
instrument, is weakened! How can we expect that when reduced to such a
degree that, as we said before, it finds new enemies in all the ordinary
difficulties of making War, it will be able to recover by fresh efforts
what has been lost! Before the battle there was a real or assumed
equilibrium between the two sides; this is lost, and, therefore, some
external assistance is requisite to restore it; every new effort without
such external support can only lead to fresh losses.

Thus, therefore, the most moderate victory of the chief Army must tend
to cause a constant sinking of the scale on the opponent's side, until
new external circumstances bring about a change. If these are not near,
if the conqueror is an eager opponent, who, thirsting for glory, pursues
great aims, then a first-rate Commander, and in the beaten Army a true
military spirit, hardened by many campaigns are required, in order to
stop the swollen stream of prosperity from bursting all bounds, and to
moderate its course by small but reiterated acts of resistance, until
the force of victory has spent itself at the goal of its career.

And now as to the effect of defeat beyond the Army, upon the Nation and
Government! It is the sudden collapse of hopes stretched to the utmost,
the downfall of all self-reliance. In place of these extinct forces,
fear, with its destructive properties of expansion, rushes into the
vacuum left, and completes the prostration. It is a real shock upon the
nerves, which one of the two athletes receives from the electric spark
of victory. And that effect, however different in its degrees, is never
completely wanting. Instead of every one hastening with a spirit of
determination to aid in repairing the disaster, every one fears that his
efforts will only be in vain, and stops, hesitating with himself, when
he should rush forward; or in despondency he lets his arm drop, leaving
everything to fate.

The consequence which this effect of victory brings forth in the course
of the War itself depend in part on the character and talent of the
victorious General, but more on the circumstances from which the victory
proceeds, and to which it leads. Without boldness and an enterprising
spirit on the part of the leader, the most brilliant victory will lead
to no great success, and its force exhausts itself all the sooner on
circumstances, if these offer a strong and stubborn opposition to it.
How very differently from Daun, Frederick the Great would have used the
victory at Kollin; and what different consequences France, in place of
Prussia, might have given a battle of Leuthen!

The conditions which allow us to expect great results from a great
victory we shall learn when we come to the subjects with which they are
connected; then it will be possible to explain the disproportion which
appears at first sight between the magnitude of a victory and its
results, and which is only too readily attributed to a want of energy
on the part of the conqueror. Here, where we have to do with the great
battle in itself, we shall merely say that the effects now depicted
never fail to attend a victory, that they mount up with the intensive
strength of the victory--mount up more the more the whole strength of
the Army has been concentrated in it, the more the whole military power
of the Nation is contained in that Army, and the State in that military
power.

On War.txt

But then the question may be asked, Can theory accept this effect of victory as absolutely necessary?--must it not rather endeavour to find out counteracting means capable of neutralising these effects? It seems quite natural to answer this question in the affirmative; but heaven defend us from taking that wrong course of most theories, out of which is begotten a mutually devouring Pro et Contra.

Certainly that effect is perfectly necessary, for it has its foundation in the nature of things, and it exists, even if we find means to struggle against it; just as the motion of a cannon ball is always in the direction of the terrestrial, although when fired from east to west part of the general velocity is destroyed by this opposite motion.

All War supposes human weakness, and against that it is directed.

Therefore, if hereafter in another place we examine what is to be done after the loss of a great battle, if we bring under review the resources which still remain, even in the most desperate cases, if we should express a belief in the possibility of retrieving all, even in such a case; it must not be supposed we mean thereby that the effects of such a defeat can by degrees be completely wiped out, for the forces and means used to repair the disaster might have been applied to the realisation of some positive object; and this applies both to the moral and physical forces.

Another question is, whether, through the loss of a great battle, forces are not perhaps roused into existence, which otherwise would never have come to life. This case is certainly conceivable, and it is what has actually occurred with many Nations. But to produce this intensified reaction is beyond the province of military art, which can only take account of it where it might be assumed as a possibility.

If there are cases in which the fruits of a victory appear rather of a destructive nature in consequence of the reaction of the forces which it had the effect of rousing into activity--cases which certainly are very exceptional--then it must the more surely be granted, that there is a difference in the effects which one and the same victory may produce according to the character of the people or state, which has been conquered.

CHAPTER XI. THE USE OF THE BATTLE (continued)

WHATEVER form the conduct of War may take in particular cases, and whatever we may have to admit in the sequel as necessary respecting it: we have only to refer to the conception of War to be convinced of what follows:

1. The destruction of the enemy's military force, is the leading principle of War, and for the whole chapter of positive action the direct way to the object.

2. This destruction of the enemy's force, must be principally effected by means of battle.

3. Only great and general battles can produce great results.

4. The results will be greatest when combats unite themselves in one great battle.

5. It is only in a great battle that the General-in-Chief commands in person, and it is in the nature of things, that he should place more confidence in himself than in his subordinates.

From these truths a double law follows, the parts of which mutually support each other; namely, that the destruction of the enemy's military force is to be sought for principally by great battles, and their results; and that the chief object of great battles must be the destruction of the enemy's military force.

No doubt the annihilation-principle is to be found more or less in other means--granted there are instances in which through favourable circumstances in a minor combat, the destruction of the enemy's forces has been disproportionately great (Maxen), and on the other hand in a battle, the taking or holding a single post may be predominant in importance as an object--but as a general rule it remains a paramount truth, that battles are only fought with a view to the destruction of the enemy's Army, and that this destruction can only be effected by their means.

The battle may therefore be regarded as War concentrated, as the centre of effort of the whole War or campaign. As the sun's rays unite in the focus of the concave mirror in a perfect image, and in the fulness of their heat; to the forces and circumstances of War, unite in a focus in the great battle for one concentrated utmost effort.

The very assemblage of forces in one great whole, which takes place more or less in all Wars, indicates an intention to strike a decisive blow with this whole, either voluntarily as assailant, or constrained by the opposite party as defender. When this great blow does not follow, then some modifying, and retarding motives have attached themselves to the original motive of hostility, and have weakened, altered or completely checked the movement. But also, even in this condition of mutual inaction which has been the key-note in so many Wars, the idea of a possible battle serves always for both parties as a point of direction, a distant focus in the construction of their plans. The more War is War in earnest, the more it is a venting of animosity and hostility, a mutual struggle to overpower, so much the more will all activities join deadly contest, and also the more prominent in importance becomes the battle.

In general, when the object aimed at is of a great and positive nature, one therefore in which the interests of the enemy are deeply concerned, the battle offers itself as the most natural means; it is, therefore, also the best as we shall show more plainly hereafter: and, as a rule, when it is evaded from aversion to the great decision, punishment follows.

The positive object belong to the offensive, and therefore the battle is also more particularly his means. But without examining the conception of offensive and defensive more minutely here, we must still observe that, even for the defender in most cases, there is no other effectual means with which to meet the exigencies of his situation, to solve the problem presented to him.

The battle is the bloodiest way of solution. True, it is not merely reciprocal slaughter, and its effect is more a killing of the enemy's courage than of the enemy's soldiers, as we shall see more plainly in the next chapter--but still blood is always its price, and slaughter its character as well as name;(*) from this the humanity in the General's mind recoils with horror.

(*) "Schlacht", from schlachten = to slaughter.

But the soul of the man trembles still more at the thought of the decision to be given with one single blow. IN ONE POINT of space and time all action is here pressed together, and at such a moment there is stirred up within us a dim feeling as if in this narrow space all our forces could not develop themselves and come into activity, as if we had already gained much by mere time, although this time owes us nothing at all. This is all mere illusion, but even as illusion it is something, and the same weakness which seizes upon the man in every other momentous decision may well be felt more powerfully by the General, when he must stake interests of such enormous weight upon one venture.

Thus, then, Statesmen and Generals have at all times endeavoured to avoid the decisive battle, seeking either to attain their aim without it, or dropping that aim unperceived. Writers on history and theory have then busied themselves to discover in some other feature in these campaigns not only an equivalent for the decision by battle which has been avoided, but even a higher art. In this way, in the present age, it came very near to this, that a battle in the economy of war was looked upon as an evil, rendered necessary through some error committed, a morbid paroxysm to which a regular prudent system of war would never lead: only those Generals were to deserve laurels who knew how to carry on war without spilling blood, and the theory of war--a real business for Brahmins--was to be specially directed to teaching this.

Contemporary history has destroyed this illusion,(*) but no one can guarantee that it will not sooner or later reproduce itself, and lead those at the head of affairs to perversities which please man's weakness, and therefore have the greater affinity for his nature. Perhaps, by-and-by, Buonaparte's campaigns and battles will be looked upon as mere acts of barbarism and stupidity, and we shall once more turn with satisfaction and confidence to the dress-sword of obsolete and musty institutions and forms. If theory gives a caution against this, then it renders a real service to those who listen to its warning voice. MAY WE SUCCEED IN LENDING A HAND TO THOSE WHO IN OUR DEAR NATIVE LAND ARE CALLED UPON TO SPEAK WITH AUTHORITY ON THESE MATTERS, THAT WE MAY BE THEIR GUIDE INTO THIS FIELD OF INQUIRY, AND EXCITE THEM TO MAKE A CANDID EXAMINATION OF THE SUBJECT.(**)

(*) On the Continent only, it still preserves full vitality in the minds of British politicians and pressmen.--EDITOR.

(**) This prayer was abundantly granted--vide the German victories of 1870.--EDITOR.

Not only the conception of war but experience also leads us to look for a great decision only in a great battle. From time immemorial, only great victories have led to great successes on the offensive side in the absolute form, on the defensive side in a manner more or less satisfactory. Even Buonaparte would not have seen the day of Ulm, unique in its kind, if he had shrunk from shedding blood; it is rather to be regarded as only a second crop from the victorious events in his preceding campaigns. It is not only bold, rash, and presumptuous Generals who have sought to complete their work by the great venture of a decisive battle, but also fortunate ones as well; and we may rest satisfied with the answer which they have thus given to this vast question.

Let us not hear of Generals who conquer without bloodshed. If a bloody slaughter is a horrible sight, then that is a ground for paying more respect to war, but not for making the sword we wear blunter and blunter by degrees from feelings of humanity, until some one steps in with one that is sharp and lops off the arm from our body.

We look upon a great battle as a principal decision, but certainly not as the only one necessary for a war or a campaign. Instances of a great battle deciding a whole campaign, have been frequent only in modern times, those which have decided a whole war, belong to the class of rare exceptions.

A decision which is brought about by a great battle depends naturally not on the battle itself, that is on the mass of combatants engaged in it, and on the intensity of the victory, but also on a number of other relations between the military forces opposed to each other, and between the States to which these forces belong. But at the same time that the principal mass of the force available is brought to the great duel, a great decision is also brought on, the extent of which may perhaps be foreseen in many respects, though not in all, and which although not the only one, still is the FIRST decision, and as such, has an influence on those which succeed. Therefore a deliberately planned great battle, according to its relations, is more or less, but always in some degree, to be regarded as the leading means and central point of the whole system. The more a General takes the field in the true spirit of War as well as of every contest, with the feeling and the idea, that is the conviction, that he must and will conquer, the more he will strive to throw every weight into the scale in the first battle, hope and strive to win everything by it. Buonaparte hardly ever entered upon a War without thinking of conquering his enemy at once in the first battle,(*) and Frederick the Great, although in a more limited sphere, and with interests of less magnitude at stake, thought the same when, at the head of a small Army, he sought to disengage his rear from the Russians or the Federal Imperial Army.

(*) This was Moltke's essential idea in his preparations for the War of 1870. See his secret memorandum issued to G.O.C.s on May 7. 1870, pointing to a battle on the Upper Saar as his primary purpose.--EDITOR.

The decision which is given by the great battle, depends, we have said, partly on the battle itself, that is on the number of troops engaged, and partly on the magnitude of the success.

How the General may increase its importance in respect to the first point is evident in itself and we shall merely observe that according to the importance of the great battle, the number of cases which are decided along with it increases, and that therefore Generals who, confident in themselves have been lovers of great decisions, have always managed to make use of the greater part of their troops in it without neglecting on that account essential points elsewhere.

As regards the consequences or speaking more correctly the effectiveness of a victory, that depends chiefly on four points:

1. On the tactical form adopted as the order of battle.

2. On the nature of the country.

3. On the relative proportions of the three arms.

4. On the relative strength of the two Armies.

A battle with parallel fronts and without any action against a flank will seldom yield as great success as one in which the defeated Army has been turned, or compelled to change front more or less. In a broken or hilly country the successes are likewise smaller, because the power of the blow is everywhere less.

If the cavalry of the vanquished is equal or superior to that of the victor, then the effects of the pursuit are diminished, and by that great part of the results of victory are lost.

Finally it is easy to understand that if superior numbers are on the side of the conqueror, and he uses his advantage in that respect to turn the flank of his adversary, or compel him to change front, greater results will follow than if the conqueror had been weaker in numbers than the vanquished. The battle of Leuthen may certainly be quoted as a practical refutation of this principle, but we beg permission for once to say what we otherwise do not like, NO RULE WITHOUT AN EXCEPTION.

In all these ways, therefore, the Commander has the means of giving his battle a decisive character; certainly he thus exposes himself to an increased amount of danger, but his whole line of action is subject to that dynamic law of the moral world.

There is then nothing in War which can be put in comparison with the great battle in point of importance, AND THE ACME OF STRATEGIC ABILITY IS DISPLAYED IN THE PROVISION OF MEANS FOR THIS GREAT EVENT, IN THE SKILFUL DETERMINATION OF PLACE AND TIME, AND DIRECTION OF TROOPS, AND

ITS THE GOOD USE MADE OF SUCCESS.

But it does not follow from the importance of these things that they must be of a very complicated and recondite nature; all is here rather simple, the art of combination by no means great; but there is great need of quickness in judging of circumstances, need of energy, steady resolution, a youthful spirit of enterprise--heroic qualities, to which we shall often have to refer. There is, therefore, but little wanted here of that which can be taught by books and there is much that, if it can be taught at all, must come to the General through some other medium than printer's type.

The impulse towards a great battle, the voluntary, sure progress to it, must proceed from a feeling of innate power and a clear sense of the necessity; in other words, it must proceed from inborn courage and from perceptions sharpened by contact with the higher interests of life.

Great examples are the best teachers, but it is certainly a misfortune if a cloud of theoretical prejudices comes between, for even the sunbeam is refracted and tinted by the clouds. To destroy such prejudices, which many a time rise and spread themselves like a miasma, is an imperative duty of theory, for the misbegotten offspring of human reason can also be in turn destroyed by pure reason.

CHAPTER XII. STRATEGIC MEANS OF UTILISING VICTORY

THE more difficult part, viz., that of perfectly preparing the victory, is a silent service of which the merit belongs to Strategy and yet for which it is hardly sufficiently commended. It appears brilliant and full of renown by turning to good account a victory gained.

What may be the special object of a battle, how it is connected with the whole system of a War, whither the career of victory may lead according to the nature of circumstances, where its culminating-point lies--all these are things which we shall not enter upon until hereafter. But under any conceivable circumstances the fact holds good, that without a pursuit no victory can have a great effect, and that, however short the career of victory may be, it must always lead beyond the first steps in pursuit; and in order to avoid the frequent repetition of this, we shall now dwell for a moment on this necessary supplement of victory in general.

The pursuit of a beaten Army commences at the moment that Army, giving up the combat, leaves its position; all previous movements in one direction and another belong not to that but to the progress of the battle itself. Usually victory at the moment here described, even if it is certain, is still as yet small and weak in its proportions, and would not rank as an event of any great positive advantage if not completed by a pursuit on the first day. Then it is mostly, as we have before said, that the trophies which give substance to the victory begin to be gathered up. Of this pursuit we shall speak in the next place.

Usually both sides come into action with their physical powers considerably deteriorated, for the movements immediately preceding have generally the character of very urgent circumstances. The efforts which the forging out of a great combat costs, complete the exhaustion; from this it follows that the victorious party is very little less disorganised and out of his original formation than the vanquished, and therefore requires time to reform, to collect stragglers, and issue fresh ammunition to those who are without. All these things place the conqueror himself in the state of crisis of which we have already spoken. If now the defeated force is only a detached portion of the enemy's Army, or if it has otherwise to expect a considerable reinforcement, then the conqueror may easily run into the obvious danger of having to pay dear for his victory, and this consideration, in such a case, very soon puts an end to pursuit, or at least restricts it

materially. Even when a strong accession of force by the enemy is not
to be feared, the conqueror finds in the above circumstances a powerful
check to the vivacity of his pursuit. There is no reason to fear
that the victory will be snatched away, but adverse combats are still
possible, and may diminish the advantages which up to the present have
been gained. Moreover, at this moment the whole weight of all that is
sensuous in an Army, its wants and weaknesses, are dependent on the will
of the Commander. All the thousands under his command require rest
and refreshment, and long to see a stop put to toil and danger for the
present; only a few, forming an exception, can see and feel beyond the
present moment, it is only amongst this little number that there is
sufficient mental vigour to think, after what is absolutely necessary at
the moment has been done, upon those results which at such a moment only
appear to the rest as mere embellishments of victory--as a luxury of
triumph. But all these thousands have a voice in the council of the
General, for through the various steps of the military hierarchy these
interests of the sensuous creature have their sure conductor into the
heart of the Commander. He himself, through mental and bodily fatigue,
is more or less weakened in his natural activity, and thus it happens
then that, mostly from these causes, purely incidental to human nature,
less is done than might have been done, and that generally what is done
is to be ascribed entirely to the THIRST FOR GLORY, the energy, indeed
also the HARD-HEARTEDNESS of the General-in-Chief. It is only thus we
can explain the hesitating manner in which many Generals follow up a
victory which superior numbers have given them. The first pursuit of the
enemy we limit in general to the extent of the first day, including the
night following the victory. At the end of that period the necessity of
rest ourselves prescribes a halt in any case.

This first pursuit has different natural degrees.

The first is, if cavalry alone are employed; in that case it amounts
usually more to alarming and watching than to pressing the enemy in
reality, because the smallest obstacle of ground is generally sufficient
to check the pursuit. Useful as cavalry may be against single bodies of
broken demoralised troops, still when opposed to the bulk of the beaten
Army it becomes again only the auxiliary arm, because the troops in
retreat can employ fresh reserves to cover the movement, and, therefore,
at the next trifling obstacle of ground, by combining all arms they can
make a stand with success. The only exception to this is in the case of
an army in actual flight in a complete state of dissolution.

The second degree is, if the pursuit is made by a strong advance-guard
composed of all arms, the greater part consisting naturally of cavalry.
Such a pursuit generally drives the enemy as far as the nearest strong
position for his rear-guard, or the next position affording space for
his Army. Neither can usually be found at once, and, therefore, the
pursuit can be carried further; generally, however, it does not extend
beyond the distance of one or at most a couple of leagues, because
otherwise the advance-guard would not feel itself sufficiently
supported. The third and most vigorous degree is when the victorious
Army itself continues to advance as far as its physical powers can
endure. In this case the beaten Army will generally quit such ordinary
positions as a country usually offers on the mere show of an attack, or
of an intention to turn its flank; and the rear-guard will be still less
likely to engage in an obstinate resistance.

In all three cases the night, if it sets in before the conclusion of
the whole act, usually puts an end to it, and the few instances in which
this has not taken place, and the pursuit has been continued throughout
the night, must be regarded as pursuits in an exceptionally vigorous
form.

If we reflect that in fighting by night everything must be, more or
less, abandoned to chance, and that at the conclusion of a battle the
regular cohesion and order of things in an army must inevitably be
disturbed, we may easily conceive the reluctance of both Generals to
carrying on their business under such disadvantageous conditions. If a

On War.txt
complete dissolution of the vanquished Army, or a rare superiority
of the victorious Army in military virtue does not ensure success,
everything would in a manner be given up to fate, which can never be for
the interest of any one, even of the most fool-hardy General. As a rule,
therefore, night puts an end to pursuit, even when the battle has only
been decided shortly before darkness sets in. This allows the conquered
either time for rest and to rally immediately, or, if he retreats
during the night it gives him a march in advance. After this break the
conquered is decidedly in a better condition; much of that which had
been thrown into confusion has been brought again into order, ammunition
has been renewed, the whole has been put into a fresh formation.
Whatever further encounter now takes place with the enemy is a new
battle not a continuation of the old, and although it may be far from
promising absolute success, still it is a fresh combat, and not merely a
gathering up of the debris by the victor.

When, therefore, the conqueror can continue the pursuit itself
throughout the night, if only with a strong advance-guard composed
of all arms of the service, the effect of the victory is immensely
increased, of this the battles of Leuthen and La Belle Alliance(*) are
examples.

    (*) Waterloo.

The whole action of this pursuit is mainly tactical, and we only dwell
upon it here in order to make plain the difference which through it may
be produced in the effect of a victory.

This first pursuit, as far as the nearest stopping-point, belongs as a
right to every conqueror, and is hardly in any way connected with his
further plans and combinations. These may considerably diminish the
positive results of a victory gained with the main body of the Army, but
they cannot make this first use of it impossible; at least cases of that
kind, if conceivable at all, must be so uncommon that they should have
no appreciable influence on theory. And here certainly we must say
that the example afforded by modern Wars opens up quite a new field for
energy. In preceding Wars, resting on a narrower basis, and altogether
more circumscribed in their scope, there were many unnecessary
conventional restrictions in various ways, but particularly in this
point. THE CONCEPTION, HONOUR OF VICTORY seemed to Generals so much
by far the chief thing that they thought the less of the complete
destruction of the enemy's military force, as in point of fact that
destruction of force appeared to them only as one of the many means in
War, not by any means as the principal, much less as the only means; so
that they the more readily put the sword in its sheath the moment the
enemy had lowered his. Nothing seemed more natural to them than to
stop the combat as soon as the decision was obtained, and to regard all
further carnage as unnecessary cruelty. Even if this false philosophy
did not determine their resolutions entirely, still it was a point
of view by which representations of the exhaustion of all powers, and
physical impossibility of continuing the struggle, obtained readier
evidence and greater weight. Certainly the sparing one's own instrument
of victory is a vital question if we only possess this one, and foresee
that soon the time may arrive when it will not be sufficient for all
that remains to be done, for every continuation of the offensive must
lead ultimately to complete exhaustion. But this calculation was still
so far false, as the further loss of forces by a continuance of the
pursuit could bear no proportion to that which the enemy must suffer.
That view, therefore, again could only exist because the military forces
were not considered the vital factor. And so we find that in former Wars
real heroes only--such as Charles XII., Marlborough, Eugene, Frederick
the Great--added a vigorous pursuit to their victories when they were
decisive enough, and that other Generals usually contented themselves
with the possession of the field of battle. In modern times the greater
energy infused into the conduct of Wars through the greater importance
of the circumstances from which they have proceeded has thrown down
these conventional barriers; the pursuit has become an all-important
business for the conqueror; trophies have on that account multiplied in
Page 153

extent, and if there are cases also in modern Warfare in which this has not been the case, still they belong to the list of exceptions, and are to be accounted for by peculiar circumstances.

At Gorschen(*) and Bautzen nothing but the superiority of the allied cavalry prevented a complete rout, at Gross Beeren and Dennewitz the ill-will of Bernadotte, the Crown Prince of Sweden; at Laon the enfeebled personal condition of Bluecher, who was then seventy years old and at the moment confined to a dark room owing to an injury to his eyes.

(*) Gorschen or Lutzen, May 2, 1813; Gross Beeren and Dennewitz, August 22, 1813; Bautzen. May 22, 1913; Laon, March 10 1813.

But Borodino is also an illustration to the point here, and we cannot resist saying a few more words about it, partly because we do not consider the circumstances are explained simply by attaching blame to Buonaparte, partly because it might appear as if this, and with it a great number of similar cases, belonged to that class which we have designated as so extremely rare, cases in which the general relations seize and fetter the General at the very beginning of the battle. French authors in particular, and great admirers of Buonaparte (Vaudancourt, Chambray, Se'gur), have blamed him decidedly because he did not drive the Russian Army completely off the field, and use his last reserves to scatter it, because then what was only a lost battle would have been a complete rout. We should be obliged to diverge too far to describe circumstantially the mutual situation of the two Armies; but this much is evident, that when Buonaparte passed the Niemen with his Army the same corps which afterwards fought at Borodino numbered 300,000 men, of whom now only 120,000 remained, he might therefore well be apprehensive that he would not have enough left to march upon Moscow, the point on which everything seemed to depend. The victory which he had just gained gave him nearly a certainty of taking that capital, for that the Russians would be in a condition to fight a second battle within eight days seemed in the highest degree improbable; and in Moscow he hoped to find peace. No doubt the complete dispersion of the Russian Army would have made this peace much more certain; but still the first consideration was to get to Moscow, that is, to get there with a force with which he should appear dictator over the capital, and through that over the Empire and the Government. The force which he brought with him to Moscow was no longer sufficient for that, as shown in the sequel, but it would have been still less so if, in scattering the Russian Army, he had scattered his own at the same time. Buonaparte was thoroughly alive to all this, and in our eyes he stands completely justified. But on that account this case is still not to be reckoned amongst those in which, through the general relations, the General is interdicted from following up his victory, for there never was in his case any question of mere pursuit. The victory was decided at four o'clock in the afternoon, but the Russians still occupied the greater part of the field of battle; they were not yet disposed to give up the ground, and if the attack had been renewed, they would still have offered a most determined resistance, which would have undoubtedly ended in their complete defeat, but would have cost the conqueror much further bloodshed. We must therefore reckon the Battle of Borodino as amongst battles, like Bautzen, left unfinished. At Bautzen the vanquished preferred to quit the field sooner; at Borodino the conqueror preferred to content himself with a half victory, not because the decision appeared doubtful, but because he was not rich enough to pay for the whole.

Returning now to our subject, the deduction from our reflections in relation to the first stage of pursuit is, that the energy thrown into it chiefly determines the value of the victory; that this pursuit is a second act of the victory, in many cases more important also than the first, and that strategy, whilst here approaching tactics to receive from it the harvest of success, exercises the first act of her authority by demanding this completion of the victory.

But further, the effects of victory are very seldom found to stop with
this first pursuit; now first begins the real career to which victory
lent velocity. This course is conditioned as we have already said, by
other relations of which it is not yet time to speak. But we must here
mention, what there is of a general character in the pursuit in order to
avoid repetition when the subject occurs again.

In the further stages of pursuit, again, we can distinguish three
degrees: the simple pursuit, a hard pursuit, and a parallel march to
intercept.

The simple FOLLOWING or PURSUING causes the enemy to continue his
retreat, until he thinks he can risk another battle. It will therefore
in its effect suffice to exhaust the advantages gained, and besides
that, all that the enemy cannot carry with him, sick, wounded, and
disabled from fatigue, quantities of baggage, and carriages of all
kinds, will fall into our hands, but this mere following does not
tend to heighten the disorder in the enemy's Army, an effect which is
produced by the two following causes.

If, for instance, instead of contenting ourselves with taking up every
day the camp the enemy has just vacated, occupying just as much of the
country as he chooses to abandon, we make our arrangements so as
every day to encroach further, and accordingly with our advance-guard
organised for the purpose, attack his rear-guard every time it attempts
to halt, then such a course will hasten his retreat, and consequently
tend to increase his disorganisation.--This it will principally effect
by the character of continuous flight, which his retreat will thus
assume. Nothing has such a depressing influence on the soldier, as the
sound of the enemy's cannon afresh at the moment when, after a forced
march he seeks some rest; if this excitement is continued from day to
day for some time, it may lead to a complete rout. There lies in it a
constant admission of being obliged to obey the law of the enemy, and of
being unfit for any resistance, and the consciousness of this cannot do
otherwise than weaken the moral of an Army in a high degree. The effect
of pressing the enemy in this way attains a maximum when it drives
the enemy to make night marches. If the conqueror scares away the
discomfited opponent at sunset from a camp which has just been taken
up either for the main body of the Army, or for the rear-guard, the
conquered must either make a night march, or alter his position in
the night, retiring further away, which is much the same thing; the
victorious party can on the other hand pass the night in quiet.

The arrangement of marches, and the choice of positions depend in this
case also upon so many other things, especially on the supply of the
Army, on strong natural obstacles in the country, on large towns,
&c. &c., that it would be ridiculous pedantry to attempt to show by a
geometrical analysis how the pursuer, being able to impose his laws on
the retreating enemy, can compel him to march at night while he takes
his rest. But nevertheless it is true and practicable that marches
in pursuit may be so planned as to have this tendency, and that the
efficacy of the pursuit is very much enchanced thereby. If this is
seldom attended to in the execution, it is because such a procedure
is more difficult for the pursuing Army, than a regular adherence to
ordinary marches in the daytime. To start in good time in the morning,
to encamp at mid-day, to occupy the rest of the day in providing for the
ordinary wants of the Army, and to use the night for repose, is a
much more convenient method than to regulate one's movements exactly
according to those of the enemy, therefore to determine nothing till the
last moment, to start on the march, sometimes in the morning, sometimes
in the evening, to be always for several hours in the presence of the
enemy, and exchanging cannon shots with him, and keeping up skirmishing
fire, to plan manoeuvres to turn him, in short, to make the whole
outlay of tactical means which such a course renders necessary. All that
naturally bears with a heavy weight on the pursuing Army, and in War,
where there are so many burdens to be borne, men are always inclined
to strip off those which do not seem absolutely necessary. These
observations are true, whether applied to a whole Army or as in the more

usual case, to a strong advance-guard. For the reasons just mentioned, this second method of pursuit, this continued pressing of the enemy pursued is rather a rare occurrence; even Buonaparte in his Russian campaign, 1812, practised it but little, for the reasons here apparent, that the difficulties and hardships of this campaign, already threatened his Army with destruction before it could reach its object; on the other hand, the French in their other campaigns have distinguished themselves by their energy in this point also.

Lastly, the third and most effectual form of pursuit is, the parallel march to the immediate object of the retreat.

Every defeated Army will naturally have behind it, at a greater or less distance, some point, the attainment of which is the first purpose in view, whether it be that failing in this its further retreat might be compromised, as in the case of a defile, or that it is important for the point itself to reach it before the enemy, as in the case of a great city, magazines, &c., or, lastly, that the Army at this point will gain new powers of defence, such as a strong position, or junction with other corps.

Now if the conqueror directs his march on this point by a lateral road, it is evident how that may quicken the retreat of the beaten Army in a destructive manner, convert it into hurry, perhaps into flight.(*) The conquered has only three ways to counteract this: the first is to throw himself in front of the enemy, in order by an unexpected attack to gain that probability of success which is lost to him in general from his position; this plainly supposes an enterprising bold General, and an excellent Army, beaten but not utterly defeated; therefore, it can only be employed by a beaten Army in very few cases.

(*) This point is exceptionally well treated by von Bernhardi in his "Cavalry in Future Wars." London: Murray, 1906.

The second way is hastening the retreat; but this is just what the conqueror wants, and it easily leads to immoderate efforts on the part of the troops, by which enormous losses are sustained, in stragglers, broken guns, and carriages of all kinds.

The third way is to make a detour, and get round the nearest point of interception, to march with more ease at a greater distance from the enemy, and thus to render the haste required less damaging. This last way is the worst of all, it generally turns out like a new debt contracted by an insolvent debtor, and leads to greater embarrassment. There are cases in which this course is advisable; others where there is nothing else left; also instances in which it has been successful; but upon the whole it is certainly true that its adoption is usually influenced less by a clear persuasion of its being the surest way of attaining the aim than by another inadmissible motive--this motive is the dread of encountering the enemy. Woe to the Commander who gives in to this! However much the moral of his Army may have deteriorated, and however well founded may be his apprehensions of being at a disadvantage in any conflict with the enemy, the evil will only be made worse by too anxiously avoiding every possible risk of collision. Buonaparte in 1813 would never have brought over the Rhine with him the 30,000 or 40,000 men who remained after the battle of Hanau,(*) if he had avoided that battle and tried to pass the Rhine at Mannheim or Coblenz. It is just by means of small combats carefully prepared and executed, and in which the defeated army being on the defensive, has always the assistance of the ground--it is just by these that the moral strength of the Army can first be resuscitated.

(*) At Hanau (October 30, 1813), the Bavarians some 50,000 strong threw themselves across the line of Napoleon's retreat from Leipsic. By a masterly use of its artillery the French tore the Bavarians asunder and marched on over their bodies.--EDITOR.

On War.txt

The beneficial effect of the smallest successes is incredible; but with most Generals the adoption of this plan implies great self-command. The other way, that of evading all encounter, appears at first so much easier, that there is a natural preference for its adoption. It is therefore usually just this system of evasion which best, promotes the view of the pursuer, and often ends with the complete downfall of the pursued; we must, however, recollect here that we are speaking of a whole Army, not of a single Division, which, having been cut off, is seeking to join the main Army by making a de'tour; in such a case circumstances are different, and success is not uncommon. But there is one condition requisite to the success of this race of two Corps for an object, which is that a Division of the pursuing army should follow by the same road which the pursued has taken, in order to pick up stragglers, and keep up the impression which the presence of the enemy never fails to make. Bluecher neglected this in his, in other respects unexceptionable, pursuit after La Belle Alliance.

Such marches tell upon the pursuer as well as the pursued, and they are not advisable if the enemy's Army rallies itself upon another considerable one; if it has a distinguished General at its head, and if its destruction is not already well prepared. But when this means can be adopted, it acts also like a great mechanical power. The losses of the beaten Army from sickness and fatigue are on such a disproportionate scale, the spirit of the Army is so weakened and lowered by the constant solicitude about impending ruin, that at last anything like a well organised stand is out of the question; every day thousands of prisoners fall into the enemy's hands without striking a blow. In such a season of complete good fortune, the conqueror need not hesitate about dividing his forces in order to draw into the vortex of destruction everything within reach of his Army, to cut off detachments, to take fortresses unprepared for defence, to occupy large towns, &c. &c. He may do anything until a new state of things arises, and the more he ventures in this way the longer will it be before that change will take place. There is no want of examples of brilliant results from grand decisive victories, and of great and vigorous pursuits in the wars of Buonaparte. We need only quote Jena 1806, Ratisbonne 1809, Leipsic 1813, and Belle-Alliance 1815.

CHAPTER XIII. RETREAT AFTER A LOST BATTLE

IN a lost battle the power of an Army is broken, the moral to a greater degree than the physical. A second battle unless fresh favourable circumstances come into play, would lead to a complete defeat, perhaps, to destruction. This is a military axiom. According to the usual course the retreat is continued up to that point where the equilibrium of forces is restored, either by reinforcements, or by the protection of strong fortresses, or by great defensive positions afforded by the country, or by a separation of the enemy's force. The magnitude of the losses sustained, the extent of the defeat, but still more the character of the enemy, will bring nearer or put off the instant of this equilibrium. How many instances may be found of a beaten Army rallied again at a short distance, without its circumstances having altered in any way since the battle. The cause of this may be traced to the moral weakness of the adversary, or to the preponderance gained in the battle not having been sufficient to make lasting impression.

To profit by this weakness or mistake of the enemy, not to yield one inch breadth more than the pressure of circumstances demands, but above all things, in order to keep up the moral forces to as advantageous a point as possible, a slow retreat, offering incessant resistance, and bold courageous counterstrokes, whenever the enemy seeks to gain any excessive advantages, are absolutely necessary. Retreats of great Generals and of Armies inured to War have always resembled the retreat of a wounded lion, such is, undoubtedly, also the best theory.

On War.txt
It is true that at the moment of quitting a dangerous position we have
often seen trifling formalities observed which caused a waste of
time, and were, therefore, attended with danger, whilst in such cases
everything depends on getting out of the place speedily. Practised
Generals reckon this maxim a very important one. But such cases must not
be confounded with a general retreat after a lost battle. Whoever
then thinks by a few rapid marches to gain a start, and more easily
to recover a firm standing, commits a great error. The first movements
should be as small as possible, and it is a maxim in general not to
suffer ourselves to be dictated to by the enemy. This maxim cannot be
followed without bloody fighting with the enemy at our heels, but the
gain is worth the sacrifice; without it we get into an accelerated pace
which soon turns into a headlong rush, and costs merely in stragglers
more men than rear-guard combats, and besides that extinguishes the last
remnants of the spirit of resistance.

A strong rear-guard composed of picked troops, commanded by the bravest
General, and supported by the whole Army at critical moments, a careful
utilisation of ground, strong ambuscades wherever the boldness of the
enemy's advance-guard, and the ground, afford opportunity; in short,
the preparation and the system of regular small battles,--these are the
means of following this principle.

The difficulties of a retreat are naturally greater or less according as
the battle has been fought under more or less favourable circumstances,
and according as it has been more or less obstinately contested. The
battle of Jena and La Belle-Alliance show how impossible anything like
a regular retreat may become, if the last man is used up against a
powerful enemy.

Now and again it has been suggested(*) to divide for the purpose
of retreating, therefore to retreat in separate divisions or even
eccentrically. Such a separation as is made merely for convenience, and
along with which concentrated action continues possible and is kept
in view, is not what we now refer to; any other kind is extremely
dangerous, contrary to the nature of the thing, and therefore a great
error. Every lost battle is a principle of weakness and disorganisation;
and the first and immediate desideratum is to concentrate, and in
concentration to recover order, courage, and confidence. The idea of
harassing the enemy by separate corps on both flanks at the moment when
he is following up his victory, is a perfect anomaly; a faint-hearted
pedant might be overawed by his enemy in that manner, and for such a
case it may answer; but where we are not sure of this failing in our
opponent it is better let alone. If the strategic relations after
a battle require that we should cover ourselves right and left by
detachments, so much must be done, as from circumstances is unavoidable,
but this fractioning must always be regarded as an evil, and we are
seldom in a state to commence it the day after the battle itself.

(*) Allusion is here made to the works of Lloyd Bullow and
others.

If Frederick the Great after the battle of Kollin,(*) and the raising of
the siege of Prague retreated in three columns that was done not out
of choice, but because the position of his forces, and the necessity of
covering Saxony, left him no alternative, Buonaparte after the battle of
Brienne,(**) sent Marmont back to the Aube, whilst he himself passed the
Seine, and turned towards Troyes; but that this did not end in disaster,
was solely owing to the circumstance that the Allies, instead of
pursuing divided their forces in like manner, turning with the one part
(Bluecher) towards the Marne, while with the other (Schwartzenberg),
from fear of being too weak, they advanced with exaggerated caution.

(*) June 19, 1757.

(**) January 30, 1814.

CHAPTER XIV. NIGHT FIGHTING

THE manner of conducting a combat at night, and what concerns the
details of its course, is a tactical subject; we only examine it here so
far as in its totality it appears as a special strategic means.

Fundamentally every night attack is only a more vehement form of
surprise. Now at the first look of the thing such an attack appears
quite pre-eminently advantageous, for we suppose the enemy to be taken
by surprise, the assailant naturally to be prepared for everything which
can happen. What an inequality! Imagination paints to itself a picture
of the most complete confusion on the one side, and on the other side
the assailant only occupied in reaping the fruits of his advantage.
Hence the constant creation of schemes for night attacks by those who
have not to lead them, and have no responsibility, whilst these attacks
seldom take place in reality.

These ideal schemes are all based on the hypothesis that the assailant
knows the arrangements of the defender because they have been made
and announced beforehand, and could not escape notice in his
reconnaissances, and inquiries; that on the other hand, the measures of
the assailant, being only taken at the moment of execution, cannot be
known to the enemy. But the last of these is not always quite the case,
and still less is the first. If we are not so near the enemy as to have
him completely under our eye, as the Austrians had Frederick the Great
before the battle of Hochkirch (1758), then all that we know of his
position must always be imperfect, as it is obtained by reconnaissances,
patrols, information from prisoners, and spies, sources on which no firm
reliance can be placed because intelligence thus obtained is always
more or less of an old date, and the position of the enemy may have
been altered in the meantime. Moreover, with the tactics and mode of
encampment of former times it was much easier than it is now to examine
the position of the enemy. A line of tents is much easier to distinguish
than a line of huts or a bivouac; and an encampment on a line of front,
fully and regularly drawn out, also easier than one of Divisions formed
in columns, the mode often used at present. We may have the ground on
which a Division bivouacs in that manner completely under our eye, and
yet not be able to arrive at any accurate idea.

But the position again is not all that we want to know the measures
which the defender may take in the course of the combat are just as
important, and do not by any means consist in mere random shots. These
measures also make night attacks more difficult in modern Wars than
formerly, because they have in these campaigns an advantage over those
already taken. In our combats the position of the defender is more
temporary than definitive, and on that account the defender is better
able to surprise his adversary with unexpected blows, than he could
formerly.(*)

(*) All these difficulties obviously become increased as the
power of the weapons in use tends to keep the combatants
further apart.--EDITOR.

Therefore what the assailant knows of the defensive previous to a night
attack, is seldom or never sufficient to supply the want of direct
observation.

But the defender has on his side another small advantage as well, which
is that he is more at home than the assailant, on the ground which forms
his position, and therefore, like the inhabitant of a room, will find
his way about it in the dark with more ease than a stranger. He knows
better where to find each part of his force, and therefore can more
readily get at it than is the case with his adversary.

From this it follows, that the assailant in a combat at night feels the
want of his eyes just as much as the defender, and that therefore, only
particular reasons can make a night attack advisable.

Now these reasons arise mostly in connection with subordinate parts of
an Army, rarely with the Army itself; it follows that a night attack
also as a rule can only take place with secondary combats, and seldom
with great battles.

We may attack a portion of the enemy's Army with a very superior force,
consequently enveloping it with a view either to take the whole, or to
inflict very severe loss on it by an unequal combat, provided that other
circumstances are in our favour. But such a scheme can never succeed
except by a great surprise, because no fractional part of the enemy's
Army would engage in such an unequal combat, but would retire instead.
But a surprise on an important scale except in rare instances in a very
close country, can only be effected at night. If therefore we wish to
gain such an advantage as this from the faulty disposition of a portion
of the enemy's Army, then we must make use of the night, at all events,
to finish the preliminary part even if the combat itself should not open
till towards daybreak. This is therefore what takes place in all the
little enterprises by night against outposts, and other small bodies,
the main point being invariably through superior numbers, and
getting round his position, to entangle him unexpectedly in such a
disadvantageous combat, that he cannot disengage himself without great
loss.

The larger the body attacked the more difficult the undertaking, because
a strong force has greater resources within itself to maintain the fight
long enough for help to arrive.

On that account the whole of the enemy's Army can never in ordinary
cases be the object of such an attack for although it has no assistance
to expect from any quarter outside itself, still, it contains within
itself sufficient means of repelling attacks from several sides
particularly in our day, when every one from the commencement is
prepared for this very usual form of attack. Whether the enemy can
attack us on several sides with success depends generally on conditions
quite different from that of its being done unexpectedly; without
entering here into the nature of these conditions, we confine ourselves
to observing, that with turning an enemy, great results, as well as
great dangers are connected; that therefore, if we set aside special
circumstances, nothing justifies it but a great superiority, just such
as we should use against a fractional part of the enemy's Army.

But the turning and surrounding a small fraction of the enemy, and
particularly in the darkness of night, is also more practicable for this
reason, that whatever we stake upon it, and however superior the force
used may be, still probably it constitutes only a limited portion of our
Army, and we can sooner stake that than the whole on the risk of a great
venture. Besides, the greater part or perhaps the whole serves as a
support and rallying-point for the portion risked, which again very much
diminishes the danger of the enterprise.

Not only the risk, but the difficulty of execution as well confines
night enterprises to small bodies. As surprise is the real essence of
them so also stealthy approach is the chief condition of execution: but
this is more easily done with small bodies than with large, and for
the columns of a whole Army is seldom practicable. For this reason such
enterprises are in general only directed against single outposts,
and can only be feasible against greater bodies if they are without
sufficient outposts, like Frederick the Great at Hochkirch.(*) This will
happen seldomer in future to Armies themselves than to minor divisions.

(*) October 14, 1758.

In recent times, when War has been carried on with so much more rapidity
and vigour, it has in consequence often happened that Armies have
encamped very close to each other, without having a very strong system
of outposts, because those circumstances have generally occurred just at
the crisis which precedes a great decision.

But then at such times the readiness for battle on both sides is also more perfect; on the other hand, in former Wars it was a frequent practice for armies to take up camps in sight of each other, when they had no other object but that of mutually holding each other in check, consequently for a longer period. How often Frederick the Great stood for weeks so near to the Austrians, that the two might have exchanged cannon shots with each other.

But these practices, certainly more favourable to night attacks, have been discontinued in later days; and armies being now no longer in regard to subsistence and requirements for encampment, such independent bodies complete in themselves, find it necessary to keep usually a day's march between themselves and the enemy. If we now keep in view especially the night attack of an army, it follows that sufficient motives for it can seldom occur, and that they fall under one or other of the following classes.

1. An unusual degree of carelessness or audacity which very rarely occurs, and when it does is compensated for by a great superiority in moral force.

2. A panic in the enemy's army, or generally such a degree of superiority in moral force on our side, that this is sufficient to supply the place of guidance in action.

3. Cutting through an enemy's army of superior force, which keeps us enveloped, because in this all depends on surprise, and the object of merely making a passage by force, allows a much greater concentration of forces.

4. Finally, in desperate cases, when our forces have such a disproportion to the enemy's, that we see no possibility of success, except through extraordinary daring.

But in all these cases there is still the condition that the enemy's army is under our eyes, and protected by no advance-guard.

As for the rest, most night combats are so conducted as to end with daylight, so that only the approach and the first attack are made under cover of darkness, because the assailant in that manner can better profit by the consequences of the state of confusion into which he throws his adversary; and combats of this description which do not commence until daybreak, in which the night therefore is only made use of to approach, are not to be counted as night combats.

Made in the USA
Lexington, KY
04 August 2017